A GREAT PERHAPS?

'To understand how centuries-old grievances can burst into violence, and the ways to resolve violent conflict while preserving the ethical foundation of democracy, I encourage everyone interested in guerrilla warfare in the modern world to read and savour this book, and to ponder its implications.'

— Ambassador Juan-Carlos Pinzon, former Minister of Defence, Colombia

'Combining organised crime, the drugs industry, and an ideologically driven insurgency, Columbia provides an important window into the future of armed conflict. This volume provides a fascinating account of this war and explains how the government have brought the guerrillas to peace talks. An essential read.'

— Professor Theo Farrell, Head of the Department of War Studies,
King's College London

'Fifteen years ago, revolutionary guerrillas threatened at the gates of Bogota. Today they are at the peace table. *A Great Perhaps?* explains the path of war and the route to peace. It should be read by all those interested in understanding and undoing complex insurgencies.'

— General Ved Malik (rtd), former Chief of Army Staff, India

'This timely book clearly explains how with the right resources and strategies, a country can turn its security problems around. As set out in this insightful study, Colombia's experience provides useful lessons for those nations faced with insurgencies: a country that in the late 1990s was on the verge of a failed state was able to change for the better.'

— Dr Anthony Bergin, Deputy Director, Australian Strategic Policy Institute

'From my own experience, Colombia's simultaneous transitions from war to peace and poverty to prosperity contain many good lessons for others. One especially important conclusion is that if Africa wants to successfully fight terrorism, it needs strong states and a shared commitment among national institutions to do so. *A Great Perhaps?* captures perfectly the strategic lessons and tactical insights of this unsung though exceptionally noteworthy campaign.'

— Pierre Buyoya, former President of Burundi; African Union special envoy to Mali

'Like Kenya, Colombia has faced a complex insurgency, fuelled by a combination of criminality, weak governance, social and economic exclusion, and porous borders. Colombia has managed these threats through strong leadership, international partnerships, buoyant economic growth, sound intelligence, and effective security forces. But more than anything, *A Great Perhaps?* shows, refreshingly, that the difference between success and failure is down to local leadership. In this regard, for Kenya, the devolution of power and resources is key.'

— Raila Odinga, Leader of the Opposition and former Prime Minister, Kenya

'Colombia shows, among many valuable lessons, how to have a good crisis. Africans, and others, should learn from this stand-out book in applying leadership and better policy to solve fundamental problems.'

— Branko Brkic, Editor of *The Daily Maverick*

DICKIE DAVIS
DAVID KILCULLEN
GREG MILLS
DAVID SPENCER

A Great Perhaps?

Colombia: Conflict and Convergence

HURST & COMPANY, LONDON

First published in the United Kingdom in 2016 by
C. Hurst & Co. (Publishers) Ltd.,
41 Great Russell Street, London, WC1B 3PL
© Dickie Davis, David Kilcullen, Greg Mills and David Spencer, 2016
All rights reserved.
Printed in the United Kingdom by Bell and Bain Ltd, Glasgow

The right of Dickie Davis, David Kilcullen, Greg Mills and David Spencer
to be identified as the authors of this publication is asserted by them in
accordance with the Copyright, Designs and Patents Act, 1988.

A Cataloguing-in-Publication data record for this book
is available from the British Library.

ISBN: 9781849046282

This book is printed using paper from registered sustainable
and managed sources.

www.hurstpublishers.com

'I go to seek a great perhaps.'

Gabriel García Márquez,
The General in His Labyrinth

CONTENTS

ACKNOWLEDGEMENTS

This book is the product of several field research trips to Colombia during 2013, 2014, and 2015, which included two study tours by international experts from Kenya, Rwanda, South Africa, Burundi, Somalia, Nigeria, Malawi, the African Union and the African Development Bank, Australia, the United States, and the United Kingdom, led by Olusegun Obasanjo and Pierre Buyoya (the former presidents of Nigeria and Burundi respectively).

These and other research visits were facilitated and generously sponsored by the Brenthurst Foundation and supported by the Ministry of Defence of the Republic of Colombia. The authors would like to express their personal appreciation to the Ministry, and especially to Minister Juan Carlos Pinzón and Vice-Ministers Diana Quintero and Jorge Bedoya for their cordiality. As accompanying officers, Andres Brochero, Colonel Jaime Ariza and Juan Sebastian Jiménez went out of their way to assist the delegations.

The opinions expressed here are, however, those of the authors alone, and should in no way reflect on the Government of Colombia or any other institution. Moreover, while the Government of Colombia assisted the delegations with local transport and accommodation, it should be noted that the authors have not received any financial remuneration from the Government of Colombia or its agencies.

Finally, the authors would like to thank Michael Dwyer and his team at Hurst & Co. for their efficiency in seeing the volume to fruition, to Alasdair Craig for his sterling efforts in finalising and editing the manuscript, to Adrian Johnson and Cathy Haenlein for their editorial assistance, and to the staff of E Oppenheimer & Son at 1 Charterhouse Street for their kind hospitality during the final production phase.

RRD, DJK, GJBM and *DES*   London, August 2015

ABOUT THE AUTHORS

Dickie Davis is the Managing Director of Nant Enterprises Ltd and an associate of the Johannesburg-based Brenthurst Foundation. He served for thirty-one years in the British Army, reaching the rank of Major General. During his military career he served extensively on operations in Afghanistan, commanding the first UK Provincial Reconstruction Team in Mazar-e-Sharif, leading the ISAF Reconstruction and Development effort and serving as Chief of Staff of Regional Command (South). He also saw extensive service in the Balkans and on numerous other worldwide deployments. He is a Vice President of the Institution of Royal Engineers, Chairman of both the Royal Engineers' Museum and the Royal Engineers Officers' Widows Society, and is Honorary Colonel of the Royal Monmouthshire Royal Engineers (Militia). A civil engineering graduate, he holds a Master's degree in defence technology and is a Fellow of the Chartered Management Institute.

David Kilcullen is a Senior Fellow in the Future of War Program at the New America Foundation, Washington D.C., and is the Founder and Chairman of Caerus Global Solutions, a research and design consultancy and Chairman of First Mile Geo, a geospatial analysis software firm. Before founding Caerus, he served twenty-four years as a light infantry officer in the Australian Army, specialising in guerrilla and unconventional warfare, and then with the US State Department, where he was chief strategist in the Counterterrorism Bureau. Thereafter he was Senior Counterinsurgency Advisor to General David Petraeus in Iraq, and Special Advisor for Counterinsurgency to Secretary of State Condoleezza Rice. He first worked in Colombia in 2009, and has visited regularly since in support of Colombian civil and security efforts to defeat the guerrillas and bring peace. In his current role he advises governments, businesses and NGOs on complex humanitarian and security challenges, and has led field teams in Iraq, Afghanistan, Somalia, Libya and elsewhere. He is the author of numerous scholarly articles and of three books: the Washington Post bestseller *The Accidental Guerrilla: Fighting Small Wars in the Midst of a Big one* (2009), *Counterinsurgency* (2010), and *Out of the Mountains: The Coming Age of the Urban Guerrilla*, which won the 2013 PROSE award for Professional and Scholarly Excellence in Government and Politics. He holds a PhD in the Politics of Insurgency from the University of New South Wales, Australia.

Greg Mills directs the Johannesburg-based Brenthurst Foundation, and is the author of the best-selling books *Why Africa is Poor: And What Africans Can Do About It* (2010) and, with Jeffrey Herbst, *Africa's Third Liberation* (2012). In 2008 he was assigned in Kigali as Strategy Advisor to the President of Rwanda. He has directed strategic advisory groups in Malawi, Mozambique, and for ISAF-HQ in Kabul and Kandahar in Afghanistan, and worked for heads of government additionally in Liberia, Lesotho, Kenya, Zambia and Zimbabwe. He holds a PhD from Lancaster University and an Honours degree in African Studies from the University of Cape Town. A member of the International Institute for Strategic Studies and Chatham House, and of the Advisory Board of the Royal United Services Institute, in 2013 he was appointed to the African Development Bank's High-Level Panel on Fragile States. The co-editor of *Victory Among People: Lessons from Countering Insurgencies and Stabilising Fragile States* (2011), with General Sir David Richards, his most recent books are *Why States Recover: Changing Walking Societies into Winning Nations, from Afghanistan to Zimbabwe* (2015) and, with Jeffrey Herbst, the best-selling *How South Africa Works: And Must Do Better* (2015). He first worked in Colombia in 2006, and has since hosted visiting African policy-makers there in 2008, 2014 and 2015.

David Spencer is Professor of Counterterrorism/Counterinsurgency at the William J Perry Center for Hemispheric Defense Studies, having earned his doctorate in Political Science from George Washington University in 2002, where he studied Latin American Politics, specialising in regional insurgency and terrorism. For the last fifteen years he has worked in a variety of positions in support of Plan Colombia. He also served as a consultant to the Salvadoran Ministry of Defence during the 1979–1992 civil war. He has worked for a number of think tanks and consulting firms such as the Center for Naval

ABOUT THE AUTHORS

Analyses (CNA) and Science Applications International Corporation (SAIC). He was raised in Latin America, living in Chile, Costa Rica, Colombia, Venezuela and Guatemala. He served in the US Army and National Guard as an Infantryman, attaining the rank of Sergeant, and was mobilised for the First Gulf War in 1990–1. In June 2011 he published the study *Colombia's Road to Recovery: Security and Governance 1982–2010* (US Government). He is also the author of *Strategy and Tactics of the Salvadoran FMLN Guerrillas: Last Battle of the Cold War, Blueprint for Future Conflicts*, and *From Vietnam to El Salvador: The Saga of the FMLN Sappers and Other Guerrilla Special Forces in Latin America*.

LIST OF ACRONYMS

AFEUR	Agrupación de Fuerzas Especiales Antiterroristas Urbanas (Urban Counter-Terrorism Special Forces Group)
AGLAN	Agrupación de Lanceros (Lancero Group)
AQIM	Al Qaeda in the Islamic Maghreb
ARENA	National Republican Alliance (El Salvador)
AUC	Autodefensas Unidas de Colombia (United Self-Defence Forces)
BACRIM	Bandas Criminales (organised criminal gangs)
BAOEA	Batallón de Operaciones Especiales de Aviación (Air Force Special Operations Battalion)
CBD	central business district
CCAI	Centro de Coordinación de Acción Integral (Centre for Coordinated Integral Action)
CEDEZOs	Centros de Desarrollo Empresarial Zonal (Zonal Business Development Centres)
CEP	circular error probable (a measure of accuracy)
CGSB	Coordinadora Guerrillera Simón Bolívar (Simón Bolívar Guerrilla Coordinating Board)
COECG	Comando Operaciones Especiales de Contraguerrillas (Counter-Guerrilla Special Operations Command)
COESE	Comando de Operaciones Especiales-Ejercito (Army Special Operations Command)
COIN	counterinsurgency
COMINTER	international commission (FARC)
CREI	Committee for Strategic Review and Innovation
DAS	Departamento Administrativo de Seguridad (Administrative Department of Security)

DMZ	demilitarised zone
DPS	Departamento para la Prosperidad Social (Department for Social Prosperity)
DSP	Política de Seguridad Democrática (Democratic Security Policy)
ELN	Ejercito de Liberación Nacional (National Liberation Army)
EPL	Ejercito Popular de Liberación (Popular Liberation Army)
EPM	Empresas Publicas de Medellín
EPP	Ejército del Pueblo Paraguayo (Paraguayan People's Army)
ETA	Euskadi Ta Askatasuna (Basque Country and Freedom, Spain)
FARC	Fuerzas Armadas Revolucionarias de Colombia (Revolutionary Armed Forces of Colombia)
FARC-EP	Fuerzas Armadas Revolucionarias de Colombia—Ejercito del Pueblo (Revolutionary Armed Forces of Colombia—People's Army)
FARCRIM	FARC dissidents
FDI	foreign direct investment
FMLN	Faribundo Marti National Liberation Front (El Salvador)
FPMR	Manuel Rodriguez Patriotic Front
GACs	Grupo Asesor del Comandante (Commander's Advisory Groups)
HVT	High-Value Target
IEDs	improvised explosive devices
IRA	Irish Republican Army
ISAF	International Security Assistance Force
JAC	Juntas de Acción Comunal (Community Action Committee)
JOEC	Joint Operational Intelligence Committee
MAS	Movimiento al Socialismo (Movement Towards Socialism, Bolivia)
MBNC	Movimiento Bolivariano por la Nueva Colombia (Bolivarian Movement for a New Colombia)
MCP	Malayan Communist Party
MPAJA	Malayan Peoples' Anti-Japanese Army
MRLA	Malayan Races Liberation Army
MRTA	Movimiento Revolucionario Túpac Amaru (Túpac Amaru Revolutionary Movement, Peru)

NCO	non-commissioned officer
NCP	National Consolidation Plan
NFO	Nueva Forma de Operar (New Mode of Operations)
ODIN	Organizaciones Delincuenciales Integradas al Narcotráfico (Criminal Organizations Integrated into Drug Trafficking)
OECD	Organisation for Economic Cooperation and Development
OEF	Operation Enduring Freedom
PC3	Partido Communista Clandestino Colombiano (Clandestine Communist Party of Colombia)
PCC	Partido Communista de Colombia (Colombian Communist Party)
PINES	Projects of National Interest
SANDF	South African National Defence Force
SIMCI	Integrated Illicit Crop Monitoring System
UACT	Unidad Administrativa para la Consolidación Territorial (Administrative Unit for Territorial Consolidation)
UP	Unión Patriótica (Patriotic Union party)
USAID	United States Agency for International Development

LIST OF FIGURES AND MAPS

FOREWORD

Olusegun Obasanjo

Many African countries share similarities with Colombia, not least the challenge of managing difficult security circumstances. Such countries have to deal with the increasing overlap between criminality and terrorism. All of us together face perhaps the most intractable challenge for any developing country: that of social and economic inclusion capable of spanning both rural and urban divides and those between the 'haves' and 'have-nots' in our societies. Such legitimate problems—and those who seek to fix them through violent, illegitimate means—are as familiar to Africans as to Colombians.

For these reasons, in 2013, 2014 and 2015, under the auspices of the Johannesburg-based Brenthurst Foundation, teams of distinguished African political leaders, civil officials and military experts visited Colombia to learn how this country has beaten back a criminal insurgency, begun to address its historical legacy of uneven development and unequal prosperity, and set its people on a path to peace.

Like Colombia, many African countries enjoyed economic growth in the twenty-first century, the continent as a whole averaging 5 per cent annually. The possibilities for collaboration and investment are clear: great African companies like SAB Miller and Anglo American have been in Colombia for years, and there is real potential for growth and for more African companies to be involved in the Colombian economy.

Our group visited most major cities, and many towns and villages across Colombia. We spoke with military and police officers, mayors and community representatives, indigenous leaders, farmers, demobilised guerrillas, civilian officials and religious elders, and saw for ourselves the progress that has been made.

We observed Colombia's peaceful, democratic 2014 presidential elections, and cheered two Colombian victories in the football World Cup, watching the matches with the bravest of the brave—soldiers who had lost arms and legs while fighting to give every Colombian a chance for peace and prosperity. We will never forget these men, or this country.

What we saw convinces us that Africa can learn directly from Colombia in many areas, including the consistent, stable management of its economy and the transformation of its security environment. Colombia shows that security is the door through which much else follows, including in the economic domain. The close fusion of its security forces—including intelligence, police and the military—and the development of special capabilities, especially in the use of air power, offers an example of how this transformation can be achieved in short order. It illustrates fundamentally the intrinsic value of local ownership in devising solutions. Equally, it exemplifies how much we can do together, in international partnership, the likes of which underpins the Colombian miracle.

This book not only explains the steps the Colombian government has taken in its search for peace and prosperity, but highlights the areas where international cooperation can extend this model and its success. I am delighted that the Brenthurst Foundation has been behind its production. I recommend it to all engaged in what I see as the responsibility of my generation: ending conflict, ensuring stability and promoting development.

Olesegun Obasanjo
Former President of the Federal Republic of Nigeria
Chairman, The Brenthurst Foundation

PREFACE

Juan Carlos Pinzón Bueno

This book is about Colombia's recent quest for stability and governance in its campaign against terrorists, insurgents and criminals. Colombians have learned many lessons in the process, which may prove valuable to others.

As minister of defence, it was my pleasure to welcome the authors on repeated field trips to every part of the country, as they sought to compile key data and develop an independent assessment of the situation after so many years of conflict. We facilitated the research team's access to remote and sometimes dangerous parts of our country, to every branch of government, and to communities in every region in order to allow the authors to develop their own view. We imposed no restrictions on what they were able to see, whom they could speak with, or where they could go. We made an honest effort to facilitate a completely independent assessment by a group of distinguished researchers, each an expert in his field, because we were determined to expose the real lessons of our experience of conflict in this century. I should make it clear, however, that the authors did not receive any financial remuneration from the Government of Colombia.

In introducing this volume, the outcome of this considerable fieldwork, I want to complement the authors' findings with a perspective of my own, based on my experience as a member of successive Colombian governments since the 1990s, on how the campaign in this country developed, and on the key lessons that I believe can be drawn from it.

A national crisis

Like many Colombians, I am reluctant to describe what we have achieved as a 'success'—this war has been going for more than fifty years and, though everyone hopes for peace, that outcome is still uncertain. What we can say, however, is that Colombia represents a positive case study, an example of the evolution of a protracted conflict with some clear military victories for the government—and there are security, economic and social statistics that testify to our substantial progress. From 1999 to 2015, a period of sixteen years, we were able to degrade the principal insurgent movement, the Fuerzas Armadas Revolucionarias de Colombia (Revolutionary Armed Forces of Colombia, FARC), to just 35 per cent of what they were at the start of that period in terms of manpower, weapons and resources. At the same time we decapitated the terrorist movement by killing or capturing its leaders, inflicting significant losses on critical leadership structures at the strategic, operational and tactical levels. These military successes brought FARC to the peace table. Today the guerrillas, including FARC, the Ejercito de Liberación Nacional (National Liberation Army, ELN), and the Ejercito Popular de Liberación (Popular Liberation Army, EPL), are not only less numerous, but much less threatening to Colombia's stability.

By way of context, it is worth remembering what conditions were like at the end of the 1990s. FARC at that time believed they were on the point of achieving their strategic goal, which was to take Bogotá through an offensive from the eastern part of the country, by sending three columns into Cundinamarca (the state where Bogotá is located) to seize major towns and then make a final push for the capital. Taking Bogotá was FARC's Plan A, and Plan B was Cali: they sent several guerrilla columns to surround the cities and, once isolated, to seize them. Frankly speaking, the guerrillas were never close to achieving these goals, but in the late 1990s they believed they were strong enough to move forward with their strategic offensive, and that was the real problem, because it was the guerrillas' general offensive that created Colombia's national crisis between 1996 and 1999.

This plan came out of FARC's Eighth Conference of 1993, and by 1996–8 they were able to deal us a series of heavy blows—in 1998 they inflicted their largest defeat on us when they destroyed a counter-guerrilla battalion of 350 men at El Billar. Eighty of our men were killed, FARC kidnapped around 100 more, and the rest of the battalion fled. At that time it was as if the entire country was being held hostage by FARC: people were kidnapped in the cities, roads were too dangerous to drive, the guerrillas were blowing up bridges and power lines and taking towns—in a period of two years they captured 350 towns, destroying everything and removing all state presence, including police and other departments. One reaction to this insecurity was the establishment of the Autodefensas Unidas de Colombia (United Self-Defence Forces, AUC), paramilitary groups that took over the northern part of the country—but this development also proved the weakness of the state and our incapacity to protect our citizens. Both the AUC and FARC were funded by criminal activities, such as kidnapping for ransom, extortion, and especially drug trafficking.

Starting from this position of real national danger, and with the survival of Colombia as a democracy at risk, we can consider the campaign since 1999 in three phases. Although this war has been fought for all five decades of FARC's existence, it was only in the late 1990s that as a nation we decided that, to really defeat the guerrillas, we needed a sustained, integrated, long-term campaign.

Recovering territorial control, 1999–2006

In the first phase, from 1999 to 2006, we were still on the strategic defensive against FARC. So our first objective was to recover territorial control over places where people were living. The goal was to regain the major cities themselves, and the towns surrounding the cities. The first operation was Operación Libertad, which freed Cundinamarca, the department that surrounds Bogotá. Next came an operation to free the eastern part of Antioquia (Medellín), which recovered control of the whole territory, and then an operation in Cerro Berlin, where we defeated the forces that FARC sent to surround the major eastern cities of the country (Bucaramanga and Cucuta). A subsequent operation aimed to recover control of Cali and the surrounding area. We have not cleared the guerrillas out completely—even today, there is still a guerrilla presence in these areas—but these operations removed any possibility of FARC achieving its strategic objective.

The effect of these successes was highly significant for the Colombian people, because people suddenly saw that their surroundings were free, that their roads were finally safe, and that we were able to protect their economic assets.

Colombia is a huge country and unfortunately its governments have historically only focused on half of the country: the half that contains the main economic infrastructure and where 95 per cent of people live. However, progress in these areas created considerable popular support after the frustrations of the 1990s and early 2000s, when President Pastrana's peace process failed and the establishment of the El Caguan demilitarised zone strengthened

FARC and put them in a position to launch their offensive. So when the state was able to launch the first phase of its campaign, from 1999 to 2006, the effect on public support was so significant that President Álvaro Uribe was re-elected in 2006. This was the first occasion in Colombian history that a president had been re-elected, largely because people were very happy about getting their lives back to a form of normality. (In 2014, President Juan Manuel Santos became the second re-elected president.) Even after all the victories the military subsequently achieved against FARC, most ordinary Colombians still think this was the major success—and this is quite understandable, given the terrible crisis that people had lived through.

The other thing that was useful at the time was the political and international isolation of FARC, which constituted a real defeat for FARC prestige. FARC came out of the peace process of 1998–2002 very isolated internationally because President Andres Pastrana was able to portray them as drug traffickers and terrorists, and this created strong political leverage for the government. As a result, we were able to create Plan Colombia together with the United States and build the political will to fight FARC. Once President Uribe became commander in chief, he defeated FARC politically: he made people see them as criminals and terrorists instead of freedom fighters. That was the key strategic consequence of this phase of the campaign.

How did FARC lose so much strategic ground in this period? In his book on Colombia, Fidel Castro wrote that FARC never really did what they should have done to win, because they felt that they could carve out and isolate the southeast of the country by kidnapping the military and seizing major towns.[1] However, they had to use a lot of their forces in guarding concentration camps in order to control their kidnapping victims and deny this huge area to the government—in tactical terms that was a mistake. From a strategic perspective, their error was very simple: they decided to try to become a regular army, and to win through fear (using crime, drug trafficking and terrorism) instead of through popular support. So when they came to Cundinamarca and to the towns surrounding the cities, they came with an attitude of arrogance, force and destruction that turned people against them. In that first phase, we were fighting on our own territory, so we had the popular support that they lacked; yet they failed to use a guerrilla approach to win support from the people due to the arrogance of their conventional military approach. This was a critical mistake.

Of the government policies that enabled this first phase, the key one was Plan Colombia. This was Pastrana's idea, later backed by Uribe, and always with the support of their minister (now president) Juan Manuel Santos. It focused on using American financial support and assets to enhance, relatively quickly, the capabilities of the armed forces and thus improve their operational performance. In tactical and capability terms Plan Colombia improved air mobility by enabling the acquisition of US Black Hawk helicopters, and bolstered our intelligence, special operations, and training capabilities. It also boosted the government's legitimacy through human rights education. We increased our manpower substantially year-by-year, giving us the opportunity to not only push the guerrillas out, but also hold the areas re-taken. Finally we were able to create the first of our joint task forces, JTF Omega, in order to launch an offensive on one of FARC's main operating areas near Macarena. This was where Plan Colombia came into its own: if we had not created JTF Omega, we would never have put FARC on the defensive.

Plan Colombia was also part of a conscious external diplomatic strategy. President Pastrana knew we needed to strengthen the armed forces, but we were not able to do that alone: we needed support. We based our appeal to the Americans on drug trafficking—we 'got onboard' with the War on Drugs—but, with FARC so heavily involved in the drugs trade, in reality it is impossible to separate this issue from the larger campaign against FARC.

This was part, however, of a long-standing discussion with the United States, going back to the early 1960s and Plan Lazo, aimed at dealing with the so-called 'independent republics' that had emerged after La Violencia. Plan Lazo was the first real attempt at an integrated civil-military initiative, but the government failed. In essence, the military had the right idea, focusing on bringing about social solutions to resolve the guerrilla conflict, with security operations in a supporting role only. But this idea was never put into action.

After nearly forty years of war with FARC, President Pastrana realised that we not only needed to strengthen the military, but that we also required an integrated civil-military effort. And Plan Colombia was conceived with this idea in mind: we created for the first time a programme of cash transfers, Familias en Acción, to address the social and political grievances behind insurgency and criminality. But Colombia, in the late 1990s, was in the middle of its worst economic crisis in eighty years, so we needed to look for outside funding. We went to the US and to the Europeans. The US responded with real support for Colombia, although, frankly speaking, it was, during the first six years of Plan Colombia, more willing to offer money for warfare than to address the socio-political causes of the insurgency. The Pentagon and State Department were able to fund military operations, but not to put money towards addressing the issues driving the conflict. But, though it began with American military funding, in the end Plan Colombia marked a major strategic vision, something that generated results far beyond just the provision of the American assets.

The other international event that had consequences for Colombia was the terrible 11 September 2001 terrorist attack on the US. The Clinton administration's support to Colombia in 1998–9 was narrowly focused on drug trafficking. But when the tragic events of 9/11 occurred, President George W. Bush launched the War on Terror. Suddenly FARC and the ELN were not just drug traffickers, they were terrorists funded by drug trafficking. We were now able to use American assets, not only to chase the drugs, but also to pursue FARC and the ELN and to fight them. This continues today.

The war on the camps, 2007–2011

After our initial success in freeing our cities from FARC encirclement and restoring safety and freedom of movement to the people, we began to plan the second phase in September 2006, and we were able to launch it in 2007.

Colombian public discourse in 2006, after President Uribe was re-elected, had a positive tone: the public were happy that we had regained control of the roads and that people had recovered their lives again. However, the truth is that we had not yet really hit FARC in its heart. They hadn't felt us, and up to that moment, more than six years after the beginning of the campaign, it felt as though a leader of FARC had never been killed in action. That was not strictly true, in fact, but no FARC leader had ever been affected by government forces in a way that had been visible to the Colombian people.

As a result, we decided to put together a campaign to maintain what we had already achieved, but to add more building blocks. Again political continuity was really important here: it was critical that in this second phase we maintained the territorial control that we had already gained. But we also put in place the High-Value Target (HVT) campaign to extend our reach to the foremost leaders of FARC. The Colombian people were ready to support offensive action against FARC, and we were now able to destroy their military assets, bomb their camps, and get to their leaders. This explains why I describe this phase as that of the 'war on the camps'—taking the war deep into FARC's heartlands.

The first FARC leader to be reached in a way that was highly visible to the Colombian public was Tomás Medina Caracas, alias 'Negro Acacio', who died in a Colombian Air Force strike in 2007. Negro Acacio was the leading drug and weapons trafficking financier for FARC's Eastern Bloc. His loss was a big blow for them. Then we got Gustavo Rueda Díaz, alias 'Martin Caballero', commander of FARC's 37th Front in Montes de Maria and a member of their general headquarters. Next came Milton Sierra Gómez, alias 'J.J', leader of FARC's urban militia, the Frente Urbano Manuel Cepeda Vargas, who was critical for FARC's Western Bloc, which operated in the country's most heavily urbanised area. After this series of blows to FARC's leadership came Luis Edgar Devia Silva, alias 'Raúl Reyes', whose death on the border with Ecuador was the first time the Colombian government had managed to kill a member of FARC's seven-person secretariat.

Naval intelligence developed the first three of these targets, while the police developed the fourth, and later we carried out major operations led by army intelligence, such as Operation Jaque and Operation Camaleón, which rescued many of those who had been kidnapped by FARC. This was really a joint effort, across army, navy and police, to create the intelligence capability that allowed us to locate the HVTs, who the air force could then pursue. This successful effort was led by then-Defence Minister Santos.

The HVT campaign meant that we were able to destroy more than just the enemy's military assets. We defeated their aura of invulnerability—FARC had always claimed their leadership could never be touched, but we proved them wrong. A particular triumph was the death of José Juvenal Velandia, alias 'Iván Ríos', head of FARC's Central Bloc and the youngest member of the FARC high command. Ríos was killed by his own security chief as a result of our having successfully encircled his column: the guerrillas were desperate, and the security chief decided to kill his own boss as a result.

Manuel Marulanda Vélez, alias 'Tirofijo', overall head of FARC, supposedly died of natural causes at around this time. However, if he did not die because of the bombing, he died because a seventy-five year-old man cannot survive being continuously on the run through the jungle. We put his column under huge pressure, and then it came out that he had died, and FARC insisted it was from natural causes. To us the cause did not matter, but rather that FARC was in retreat.

Politically, our goal on the one hand was to make FARC feel that they were vulnerable, and on the other hand to show our civilian population that FARC was being pushed back. We were trying to push FARC away from their strategic goal of overthrowing the state through force, towards a political solution.

The objective for the Colombian government at all times has been a political solution: since the very first day of the campaign, from Pastrana, through

Uribe's time to President Santos. That continuity of political vision has been a critical part of our commitment. The government has never believed that we were going to kill every FARC member in the country. Defeating FARC's assumption of invulnerability, destroying their assets and, ultimately, the HVT campaign, brought the people a sense of victory.

In the first phase, people got a sense that they were recovering their lives, but in the second phase, there was a sense of victory—a sense that we were not just defending our cities from FARC but were going after them at their strongest points. FARC, rather than the Colombian government, were now on the defensive. What created this sense of victory was not just removing FARC from our territories, but also seeing their leaders fall, and that was very important.

Because of this, the demobilisation programme picked up steam. We also had to make very tough decisions that became very unpopular in the armed forces, but which were necessary: decisions based on restoring our legitimacy and record on human rights. We had discovered that, unfortunately, in the period between 2004 and 2008, there were a lot of killings—probably some of them of guerrillas out of combat, but certainly also some of civilians—that were presented as combat results. These so-called 'false positives' (extrajudicial killings) were a real nightmare, and our decision to prosecute the soldiers involved was extremely unpopular, but such decisions ultimately helped restore legitimacy to the armed forces, and contributed to increasing their credibility among the Colombian people. This was necessary, and ultimately it worked: in the past four years there have not been any instances of 'false positives', a fact that has been supported independently by the United Nations in their annual reports on human rights.

The final important development during this second phase was that when the Colombian people felt this sense of victory they decided to come out onto the streets to hold massive demonstrations calling for an end to FARC. The protest, on 4 February 2008, was called 'No más FARC'. This was a real political blow for FARC, one they will probably never recover from as long as they keep the brand 'FARC'. Of course, the government capitalised on what was a spontaneous movement of the people, but this movement did not originate inside the government.

In terms of the capabilities and policies that underpinned this second phase, the wealth tax (covered in Chapter 5 of this volume) was a substantial policy initiative. This tax allowed us, firstly, to give confidence to the Americans, because they saw that they were not our only source of funding. The Colombian people were not only sacrificing their own blood, but were also putting their own money towards the fight against FARC, enhancing and taking ownership over what the Americans were bringing in. Secondly, the wealth tax gave confidence to the armed forces, because they now had money for operations. The consequence of prosperity is often mismanagement or spending money where you don't need to, but having the proper resources is good for morale, commitment and confidence. The wealth tax generated that confidence, and it enhanced capabilities. We were able to buy more helicopters, to modernise some of the old assets and to buy new ones. To coin a phrase, the armed forces perceived a 'war dividend'.

We focused the build-up on capabilities that we needed. Any army will always ask for tanks, an air force will ask for fighter jets and a navy will ask for aircraft carriers, but we did not seek to acquire any such capital assets. Instead we focused on capabilities that really were able to improve what we already possessed. This is an important lesson for anyone who finds themselves in the situation we were in, for example some countries in Africa: you have to plan in detail, and develop capabilities that are directly related to your objectives. This was our approach. There were two main points of focus for this second phase of capability development. One was intelligence: we enhanced intelligence technology, improved our organisation, and the pride of President Santos is that he was able to put that intelligence capability to work, in a joint manner. We really made an effort to enhance cooperation between services and agencies, in both the intelligence and operational spheres.

The second key capability was special operations. We created a joint special operations command, putting together army, navy, air force and police units inside one joint command that was dedicated to HVTs, and this was a very substantial and useful step.

Finally, at this point in time we acquired advanced technology—including imagery and signals technology for intelligence. The intelligence technology certainly enhanced our performance, but the acquisition and deployment of precision munitions was a game-changer.

Invading the enemy's territory, 2011–2015

The context for the third phase was a rebirth, or renaissance, of FARC. We had succeeded in degrading FARC from 1999 to 2009, year by year, but suddenly from 2009 to 2010 they were able to increase their numbers, and then in 2011 they expanded again. The reason was that when Manuel Marulanda, overall head of FARC, died in 2008, Alfonso Cano took over.

By the end of 2008 Cano had created a plan to renew FARC—Plan Renacer—and the essence of this plan was to drop back a stage from the 'war of movement' (the semi-conventional approach FARC had adopted) to return to guerrilla warfare. Cano argued: 'We can no longer fight the army as if FARC is a conventional army—they are destroying us'. Cano's analysis was that FARC was committing several mistakes. First, FARC had lost popular support, without which it is impossible for a guerrilla to exist. FARC had to go back to the masses, return to getting close to the people, to infiltrate unions and social movements. And, actually, since 2008 FARC has maintained that approach—and still applies this methodology today.

Secondly, Cano argued that it was a mistake for FARC to fight the Colombian army—instead, the guerrillas were going to use IEDs as a preferred weapon, adopting a strategic offensive but implementing this through a tactical defensive. FARC would also use snipers and would not seek to fight a war of large combat units. Instead it would go back to guerrilla units, becoming a mobile guerrilla force, not attempting to hold territory but constantly moving, this being the best way to control the population. This became Cano's new approach. It was clearly a plan that focused on terrorist activities and that would lead to violations of international humanitarian law.

This forced us to rethink our approach. By 2011, when I came in as minister of defence, our problem was how to deal with FARC's Plan Renacer. We were forced to confront our assumptions from earlier in the campaign. What was our objective during Phase II? Some had thought that by killing FARC's leadership we were going to make them surrender. This idea is not unique to Colombia, of course: it is a typical assumption of counterinsurgency (COIN) warfare. But we learned that this is not sufficient. When Plan Renacer began in late 2008, there was such confidence in the HVT campaign that FARC was underestimated.

Part of this new FARC strategy was to change to a more offensive urban posture. They renewed their terrorist campaign, which they had previously ended in 2006. They were running mobile guerrilla operations in the jungle, but mounting urban asymmetric operations—bombings, assassinations and strikes against infrastructure—because they had learned the positioning of our troops, and realised that we could never secure, for example, an entire pipeline. This was not lost on the public, and the effects of FARC's tactics were amplified by the public and political criticism the Santos administration received.

This created a problem for President Santos. Not only were we struggling to provide security, but also there was the need for an endgame strategy consistent with a peace process. This became Santos' focus. At this time I was his chief of staff; Santos then asked me to be the minister of defence. Right from the outset, we knew that it was necessary to re-engage in planning: as the old General Eisenhower saying goes, 'plans are nothing, planning is everything'. We needed to really see what was going on, to understand and to put together a new campaign plan. We created the Committee for Strategic Review and Innovation (CREI) to support this planning effort. We needed to collate every bit of information on FARC base areas. While we had defeated FARC's military structure—they will never again be what they were in military terms—we had left their political structures, logistical systems, and terrorist support networks in the cities and towns virtually untouched. Now we went after those structures. To do that we focused on integration, having new judicial police units ('Groic') embedded within army units, and the Fiscalia (Attorney General's Office) working directly with the joint task forces. We came out with two plans: Sword of Honour, for fighting FARC as a system, beyond its military structure; and Green Heart, to confront the new challenges of citizen security, typical of post-conflict environments, a reality in more than 90 per cent of the municipalities of Colombia.

Twelve JTFs were created to take over the historical base areas of FARC, the ELN and the Bandas Criminales (organised criminal gangs, BACRIM): we would now become the 'invaders'. Our operations were aimed, first, at maintaining territorial control, since if security were to fail in one of the major cities—in Bogotá, Medellín, Cali or their surroundings—we would be in trouble. We had to be prepared even to over-extend ourselves in order to provide effective security. And second, we needed to maintain pressure on FARC's leadership through the HVT campaign.

On top of these existing programmes we added new aspects: dynamic planning; the fight to cut off FARC's support networks and means of finance; and occupying FARC territory. These, coupled with the increased activities of FARC, are why the level of combat in 2014 became even higher than it was in 2012 and 2013. We launched operations to affect FARC's public support in their local areas by using the military engineers' capabilities for reconstruction and infrastructure development, and we implemented the concept of GACs (Grupo Asesor del Comandante, Commander's Advisory Groups) embedded in headquarters to bring in other agencies of the state to coordinate with the military and to strengthen the bonds of the government with communities. And in the HVT campaign we struck the final blows for peace.

The consequence of this campaign was that we killed Cano, the top leader, and killed or captured another fifty-four FARC all-level commanders, the highest number ever, so that they were very significantly degraded. The leadership of BACRIM and the ELN were also affected more than ever before. FARC and ELN demobilisation rates increased again in 2013 and 2014. The capture of FARC militias rose substantially. FARC armed manpower decreased by almost 30 per cent from 2011 to 2015. The homicide rate fell to its lowest in thirty-five years, and the kidnapping rate to its lowest in more than a decade. Almost every security statistic improved, and only 6 per cent of municipalities were affected by terrorism by the end of 2014 (compared to 50 per cent fifteen years previously, and 20 per cent five years previously).

In this final phase, we have focused on creating a concept of legitimacy that goes beyond (but includes) the issue of human rights, because the concept of human rights on its own was linked to trauma for the military (because of the 'false positives' and our strenuous efforts to address that problem). It should not be a trauma any more: human rights represent the minimum baseline of legitimacy, but our legitimacy goes beyond that. For soldiers, it has to be based on the fact that the country loves you, that you are the one taking solutions to the people.

In policy and capability terms, to support this third phase of the campaign we continued the wealth tax for the same reasons as in the previous phase. This allowed us to improve the assets of the security forces, leaving the Colombian military and police at probably the peak of their historical capabilities. Second, we invested in the education of the military and police, preparing for the future by maximising our human capital, though we do not yet know what the impact of the peace process will be. Thirdly, we began working towards what we call 'legal security' for the armed forces. We paid a high price for this with NGOs and with human rights organisations, because it involved strengthening the military justice system and guaranteeing fair trials for members of the armed forces in the ordinary justice system. Of course that was unpopular with some sectors of public opinion. We always looked for reforms that would enhance the respect for human rights, improve the quality of investigations and improve discipline, but we also sought to ensure that investigators had operational knowledge so that they could gain all of the relevant facts. We based our investigations and military justice system on international humanitarian law, which was introduced to our constitution in order to guarantee the highest standards. We also created the technical defence fund, to allow members of the armed forces who were under investigation access to professional lawyers.

In terms of capabilities, we have further enhanced our domestic and regional intelligence abilities, as well as our special operations capacity. These new capabilities are being used not only for the HVT campaign, but also to fight various kinds of crimes and sources of funding for criminals. Previously we only had the special operations brigade and the joint special operations command. Now, in every division we have a battalion-level special operations unit that is trained and has the capabilities of special operations forces, and this has enhanced the reach of our higher-tier special operators. The reason that we have more HVTs killed in action today is that we now have more strike assets available. Low-intensity terrorist activity has been seen in less than 10 per cent of municipalities in response to the presence of JTFs in the guerrillas' base areas (that is, the heart of enemy territory) and as a political tool used by the guerrillas in the midst of the peace talks.

Moving the country forward towards 'normality', and to a more peaceful environment, has been a clear achievement of this period. Looking ahead, it will be critical to continue to plan and build capabilities for both current and future challenges. That will be the legacy of this period in the campaign.

Recognising victory

How can we tell if all of this ultimately adds up to a victory? After all, this conflict has been running for five decades and this is far from our first peace process. I believe that we can claim victory, because this fifteen-year-long campaign has set the conditions for a negotiated peace, which was always our national objective, and in such a way that we are negotiating from a position of strength. The security campaign was always aiming for a credible peace. War, for us, was never anything more than the means to a higher end: peace.

President Santos' decision to move towards a peace process was the result of his vision of Colombia ending its conflict of more than fifty years as the best outcome for the nation. He supported this political aim through becoming recognised as a strategist who has delivered major blows to FARC, and through his conviction that the balance of power and public support is in our side. He has applied the principle of 'nothing is agreed until everything is agreed'. He has defined four 'red lines' as the driving criteria for negotiation in a way that guarantees the victory of the nation over crime and terrorism, but also guarantees an exit with dignity for armed groups.

The first 'red line' is that we are not negotiating Colombia's economic or political model: private property, democracy, or any other aspect of the constitutional order in Colombia are not up for discussion.

Second, the president has decided not to offer a bilateral ceasefire until all aspects of the peace agreement are signed. We have thereby avoided relieving the pressure on FARC that is necessary to ensure their commitment to peace, and have ensured the exercise of the rule of law and the continuation of the fight against FARC's criminal activities (such as extortion, drug-trafficking and illegal mining) in defence of citizens' rights and public safety. The recent unilateral ceasefire from FARC is seen as a signal of their desire for peace and their inability to resist the pressure of Colombia's military.

Third, the future of the armed forces and national police is not up for negotiation. We are retaining control over the status and future of the armed forces. The peace agreements in El Salvador and Guatemala, after their civil wars, involved the elimination of their national police forces, the creation of civilian police forces and a reduction of 70 per cent in their national armed forces. In Colombia, we are not discussing such matters. We have a concept of modernisation and strengthening of the armed forces, using scenario- and capability-based planning. The government alone is dictating the future of the armed forces.

A fourth red line is the implementation of transitional justice for members of the armed forces who are under prosecution due to events during the conflict. President Santos offered a solution to members of the armed forces who committed crimes as part of the conflict that is symmetrical, but not identical, to the solution being offered to FARC. Transitional justice, in this sense, is what is feasible as defined under the standards of the Rome Statute of the International Criminal Court, and will apply even-handedly to all those affected by the conflict. This is a substantial difference from all previous peace processes in Colombia, where benefits were only granted to members of illegal groups, and not to state agents implicated in crimes related to the conflict.

In addition, a key negotiating position relates to defining the process of Disarmament, Demobilisation and Reintegration (DDR). The government retains the right to define what DDR means; for instance, in the case of demining (which is happening currently), our forces are going to be the ones who do the demining.

There are some challenges still to come. If we allow President Santos' 'red lines' to be crossed, our strategy might be weakened. This explains his strong commitment to a framework that has scope for flexibility and pragmatism, but at the same time outlines clear boundaries. Also, any failure to push FARC and the ELN to separate their criminal economies from their political ambitions will negatively impact on security.

What next for the military? In simple terms, we need to keep the edge in security to guarantee peace and defeat crime. Crime is going to happen, it will continue to exist in Colombia whatever the outcome of the peace process, so we need to keep the advantage, adapt to future threats, and always be there to protect communities. Second, we need to continue to occupy Colombia's territory and guarantee the rule of law, but in a different way in the future: no longer with offensive operations, but with engineers contributing to development, helping the population deal with natural disasters, and helping protect the environment (water sources, biodiversity and tropical forest). Third, we need to develop international cooperation, not only because this creates opportunities and constructive activity for the military, but because it builds influence for Colombia and allows us to perform a key role promoting peace in our own region and internationally as an example of a successful security model. We have already become security exporters by training more than 24,000 police and military personnel in more than sixty nations in the past four years and we have signed agreements with the EU, UN and NATO for expanding our cooperation initiative. Colombia may be one of the middle-level powers in the region because of its size, population and natural resources, but it can maintain a greater influence now because of the reputation and confidence of its armed forces and our progress in security.

Lessons for Colombia and for others

Finally, there are four general lessons for others to take from Colombia, beyond the specific details of the operations.

The first is ownership. You have to own your problem if you are going to own the solution. Colombians recognised that we had a national crisis, that others—like the US—were willing to help us with, but we first had to help ourselves, to take ownership of the problem, and only then did we begin to turn things around. It is important to highlight the resilience and capacity to overcome obstacles as a nation that Colombia has demonstrated.

The second key lesson is leadership. Three presidents in a row, over five different presidential terms, have each committed to security, and contributed policies and made decisions to bring to Colombia the rule of law and establish the country on a path to peace. The influence of all three leaders can be seen in our campaign planning—in the first phase the leadership of Pastrana and Uribe was key; in the second that of Uribe and Santos; and between the second and third phases that of Santos. These three leaders have maintained a consistent policy of trying to achieve a stable peace. President Santos has shown courage and transformational leadership skills by pushing his political capital to achieve a negotiated peace agreement under conditions of permanent political pressure.

The third lesson is commitment. There has been strong, sustained commitment from the military, from the people and from the government alike. There is a Clausewitzian 'trinity' here—government, army, people—and everyone in this relationship is committed to a positive outcome. This creates some practical realities, the first of which is adequate and sustained funding. We maintained the wealth tax, and so we had the resources available to create the capability we needed. The second reality is popular support—the military has the highest polling numbers in terms of public support that it has ever had. Institutions like the military and police are the top institutions on every poll.

Finally, of course, heroism and sacrifice: these have made a critical difference in our case. There are armies that are a lot better equipped than ours, but where we have excelled has been in our professional attitude, and more specifically in the heroism and the will to fight that we continue to have. This is intangible, but it is hugely important and arguably has proven to be the crucial differential in this war between failure and success.

Every country is different, and it's not for Colombia to preach to others. But if our experience suggests one thing above all, it's that if you plan and execute, you can change reality.

I love my brave country, and I know that with the will of the nation, leadership and the heroism and professionalism of our armed forces, Colombia will surpass any obstacle to achieve peace and reach prosperity for all in a way that will make us have a place and duty in the region and in the world.

It would be unusual if I agreed with everything the authors have written in the chapters that follow, or endorsed their judgment on every issue. Yet, to understand how centuries-old grievances can burst into violence, and the ways to resolve violent conflict while preserving the ethical foundation of democracy, I encourage everyone interested in guerrilla warfare in the modern world to read and savour this book, and to ponder its implications.

1. Map of Colombia

INTRODUCTION

COLOMBIA'S TRANSITION

David Kilcullen and Greg Mills

Such is the progress that has been made in Colombia that it is difficult to remember that, only twenty years ago, the country was famous not for its practical people or its wonderful cities and rainforests, but for its cocaine-fuelled murder rate. At the height of the drug war in the 1990s, Colombians suffered ten kidnappings a day, seventy-five political assassinations a week, and 36,000 murders a year (fifteen times the rate in the United States).[1] The military and police competed with an array of guerrillas, gangs, *narcos* and paramilitaries. Guerrillas had so isolated the largest cities that urban-dwellers travelling as few as 5 miles out of town risked kidnapping, or worse.

Colombia entered the twenty-first century at risk of becoming a failed state. Since then, national leaders have turned the situation around, applying a well-designed strategy with growing public and international support. Numbers of kidnappings and murders are down, government control has expanded, and the economy is recovering. Talks in Havana with the main insurgent group, the Fuerzas Armadas Revolucionarias de Colombia (Revolutionary Armed Forces of Colombia, FARC) offer hope of peace, even as fighting continues on the ground in key areas, and other illegal armed groups—including the Ejercito de Liberación Nacional (National Liberation Army, ELN) and a slew of right-wing paramilitaries turned organised criminal

1

gangs (Bandas Criminales, BACRIM) remain active and at large. At the same time, cocaine production and trafficking in Colombia, part of a wider hemispheric narcotics economy that shows no sign of abating, continues apace.

For many observers in the United States, Britain and elsewhere, the popular narrative has shifted dramatically over the past decade, with Colombia now seen less as a basket case and more as a shining success story. But the situation on the ground is, in truth, far shakier than it seems—and indeed, the very success of Colombia's current campaign carries the risk of future conflict.

This book, drawing on the authors' independent fieldwork in Colombia over several years, is an attempt to survey the situation as it stands in 2015, adding detail and context to that popular narrative, partly in order to correct the record and partly to help outsiders understand what has happened, and what may happen next, in Latin America's longest-running insurgent conflict. More broadly, the book's purpose is to understand the dynamics of the Colombian conflict, and distil them into observations sufficiently general and enduring that they may be of use to others facing analogous problems in other countries.

In this opening chapter, we examine Colombia's turnaround, explore current issues, and offer insights for the future and for others facing similar challenges. We consider the conflict's political economy—by which we mean the dynamic social-political-economic incentive structure that frames people's choices within a society shaped by two generations of war. One key finding is that, with some significant exceptions, key FARC commanders and others have become what we call 'conflict entrepreneurs' who seek to perpetuate war for personal gain rather to win (and thereby end) the conflict in order to achieve external objectives. Therefore, remarkable though it is, contemporary military progress will not be enough to end the war in a way that guarantees Colombia's future. A comprehensive conflict transformation is needed—one that moves Colombia from a political economy of violent exploitation, to one of inclusive, sustainable peace.

Such a transformation, as we explain in detail in later chapters, will involve not only continued military and diplomatic efforts to resolve the conflict, but also a sustained programme—over at least the next fifteen to twenty years—of economic, social and political reform, aimed at extending social and political participation to Colombia's traditionally marginalised population groups, ensuring their equitable inclusion in a functioning, licit economy, and extending state presence and the rule of law into historically under-governed areas. If this sounds like an extremely tall order, that's because it is—and maintaining the political will to sustain it will be, perhaps, the most difficult challenge of all.

Background

From Spain's conquest of the region that now includes Colombia in the 1500s, through resistance to colonialism in the eighteenth century, to the liberation wars of Simón Bolívar of 1812–19, the area has seen near-continuous conflict. Colombia is the oldest democracy in Latin America, but has been at war for 150 of its 195 years of independence: there were nine civil wars and more than fifty insurrections in the nineteenth century alone. As a result, Colombians have learned to live with 'democratic insecurity'.

Historically, conflict arose between Liberals and Conservatives—political blocs that mirrored a stratified, segmented society of European oligarchs controlling factories and huge estates, excluded rural and urban poor, and marginalised Indian and Afro-Colombian minorities. Although, for most of Colombia's history, these minorities have represented only a small fraction of Colombia's population (roughly 10 per cent in the case of Afro-Colombians and less than 4 per cent in that of indigenous Colombians) they are very unevenly distributed, with heavy concentrations of Afro-Colombians on the country's isolated Pacific coast, and indigenous populations clustered in several strategically important mountainous regions. In broader terms, the exclusion of relatively small and powerless minorities—including not only these ethnic minorities but also poor rural peasants (*campesinos*) and peri-urban workers struggling to make a living on the margins of Colombia's cities—has been an enduring source of conflict in the country.

In effect, Colombia's temperate, urbanised, heavily populated, and developed centre has historically contrasted with its tropical, rural, sparsely inhabited, and neglected periphery. Structural inequality and lack of opportunity created fertile ground for both revolutionaries seeking to overthrow the system and those who would live outside the law.

Ironically, today's conflict arose from the peace-making process after Colombia's bloodiest episode of social conflict, La Violencia, which left 300,000 dead between 1948 and 1953, when General Gustavo Rojas Pinilla took power at the request of civilian political elites, in Colombia's only twentieth-century coup d'etat, and—although ending the violence through large-scale military operations against illegal armed groups that included the forerunners of FARC—imposed military rule which lasted until 1958. The origin of the violence lay in urban elite politics: the trigger, in 1948, was the murder of Jorge Eliécer Gaitán, a Liberal Party leader, which unleashed savage violence between Liberal and Conservative militias in a civil war Colombians still remember with horror. But despite its origins, the conflict rapidly expanded beyond its original causes, and tore the country's social fabric apart. Although it derived, at least initially, from conflicts among Colombia's political elites in the cities, the violence fell most heavily on rural and small-town communities, where partisan violence among local groups was often sponsored by outside (principally urban, elite) actors. This pattern of violent clientelism ended only in 1953 when both parties, recognising they were powerless to stop the violence they had unleashed, asked General Rojas to step in to end the conflict. The period of military rule that followed Rojas's bloodless seizure of power was the army's sole twentieth-century intervention in politics, with the military acting more or less as a *deus ex machina* to reset a Colombian political system that had escaped the control of civilian elites—but Rojas's offensives against the armed groups, and his attempts to perpetuate his rule, led to mass dissatisfaction with his government, prompting a general strike in 1957 and Rojas's forced abdication by a coalition of students, civilian political parties, and dissident (that is, pro-democratic) members of the military. The end result was a political settlement, brokered by this coalition in July 1957 and validated by a national referendum on 1 December 1957, in which Liberals and Conservatives agreed to share power, alternating at the head of bipartisan National Front governments. The National Front arrangement lasted for the next sixteen years, from 1958 to 1974.

This cosy reconciliation among elites—which, by definition, excluded the poor, rural, and indigenous workers and peasants who had been most heavily affected by the violence—served the enormously beneficial purpose of ending the bloodbath of La Violencia and creating a pathway for Colombia out of military rule and back to civil democracy, but at the same time, by its very urbanised and elite nature, the National Front arrangement created the conditions for future conflict. In particular, the deal excluded communist armed movements, as well as more moderate left-wing groups, that had broken away from the traditional Liberal–Conservative dichotomy as a result of the violence. The Communist Party, in particular, refused to join the National Front process; several communist militias refused to disarm, instead establishing autonomous zones (which the central government called 'independent republics') in defiance of the peace settlement.[2] Because these 'republics'—and the armed groups controlling them—rejected the National Front arrangement, successive Colombian national unity governments (both Liberal Party- and Conservative Party-led) saw them as a threat, and as a potential trigger for the collapse of the entire 1957 peace deal and the return of massive violence. Conflict indeed began to intensify after 1959 as part of a region-wide rise in unrest after the Cuban Revolution—rural violence, for example, rose 30 per cent in 1960–62.[3] From 1959, with help from US special warfare teams and civil agencies, the Colombian government improved its counterinsurgency (COIN) capabilities, developed Plan Lazo (a comprehensive internal defence strategy), and sought to suppress the independent republics.[4] Evidence in late 1963 that Colombian guerrillas had received weapons and training from Havana underlined the regional dynamic, and prompted government action against the so-called independent republics.

In May 1964, the Colombian armed forces attacked the 'Marquetalía Republic' led by Pedro Antonio Marín, alias 'Manuel Marulanda Velez'—a charismatic leftist guerrilla leader known more widely by his nickname, Tirofijo ('Sure-shot'). The assault pushed Marulanda's guerrillas into the neighbouring area of Rio Chiquito where, in July 1964, a confederation of guerrilla groups formed the 'Southern Bloc'. Declaring themselves 'victims of the policy of fire and sword proclaimed and carried out by the oligarchic usurpers of power', the new coalition called for 'armed revolutionary struggle to win power'[5] and renamed itself the 'Fuerzas Armadas Revolucionarias de Colombia' (FARC) two years later. Also opposing the government was the rural Ejercito de Liberación Nacional (National Liberation Army, ELN), a Marxist-Leninist group espousing revolutionary theories that included Liberation Theology and a radical social justice agenda, the Maoist-oriented Ejercito Popular de Liberación (Popular Liberation Army, EPL) and, a decade later, the urban Movimiento 19 de Abril (Nineteenth of April Movement, M-19) terrorist group. The army, in turn, received support—sometimes helpful, often unwanted or embarrassing—from right-wing paramilitaries that had formed to defend communities (and wealthy landowners) threatened by the guerrillas.

The result was a long-standing and widespread guerrilla conflict defined by a domestic 'dynamic of stalemate' (discussed in detail in Chapter 3) and by the twin regional dynamics of the growing Latin American narcotics economy of the 1970s and 1980s, and the East-West ideological conflict of the Cold War. With the end of the Cold War, M-19 demobilised in 1990 and transformed itself into a parliamentary political party, but FARC and the ELN opted to continue the armed struggle, alongside (and often in competition with) the

EPL. Well before 1990, FARC had already turned to criminal activity—principally narcotics—to fund its struggle. This is not to say that FARC was non-ideological, or that FARC leaders would have been averse to greater assistance from the communist bloc, but in fact the movement never received the large-scale external support it wanted. Fidel Castro, ruling Cuba after 1959 and seeking to export a Cuban model of revolution that would 'continue making [Cuba] the example that can convert the Cordillera of the Andes into the Sierra Maestra of the American continent', trained thousands of leftist guerrillas from across the region, including Colombians.[6] But Castro distrusted orthodox communist party organisations of the sort that dominated many Latin American countries, and the Soviets provided only political, rather than material support, so that from the outset FARC was forced to be primarily self-supporting.[7] When the movement's leaders discovered, and subsequently moved into, the immensely lucrative drug economy in the early 1980s, they initially intended narcotics to be a temporary means of financing to fill the gap until external support could be found. But the sheer volume of money involved eventually seduced them, so that by the early 1990s they were fully enmeshed in the drug business as narco-guerrillas.

With the collapse of world communism in the early 1990s, the prospect of external support receded even further, and narcotics became a key source of finance, along with kidnapping and extortion. Drugs brought in an estimated $3.5 billion annually by 2005, amounting to 45 per cent of FARC's funding.[8] Involvement in the drug trade was not, of course, restricted to left-wing groups like FARC. Right-wing paramilitaries, likewise, by 2005 received three-quarters of their income from drug cartels, to which they hired out their services. This created a huge overlap between guerrillas and gangsters in Colombia.

As a consequence, FARC has evolved into a criminal-insurgent hybrid: the system it spreads to areas under its control creates its own exploitative, violent political economy in which left-wing revolutionary ideology provides a veneer for a system of racketeering built on drugs, illegal mining, extortion, robbery, and kidnapping. Colombia's insurgency has merged with criminality while FARC leaders (among others) have emerged as conflict entrepreneurs—they have discovered the value of crime as an enabler for their pursuit of raw political power.

Ideologically motivated insurgents fight for objectives extrinsic to conflict; they stop fighting when those objectives are achieved. States operate the same way: as Colombia's then defence minister Juan Carlos Pinzón observed, 'governments don't fight wars just to fight—they fight to build a better reality for their people'.[9] By contrast, conflict entrepreneurs are fighting to perpetuate a conflict, since its existence creates wealth, power and status for them: their goals are intrinsic to war.[10]

Formally, we might define a conflict entrepreneur as

> any group or individual whose profits [or, more broadly, whose power, prestige and livelihood] depend on conditions that promote conflict. Most often used to describe those who engage in or directly benefit from illegal economic activity that promotes violence or undermines efforts for good governance and economic development. These actors can exist inside or outside of government.[11]

When their stated political objectives cease to help maintain a profitable conflict, conflict entrepreneurs simply change the objectives and continue the conflict. FARC, like BACRIM, which emerged from the paramilitaries, is a classic example of this war-as-racketeering phenomenon, but it is not the only one. Many African conflicts, in particular, show a similar pattern—including clan warfare in Somalia, conflicts in Sudan and the Congo, and the Lord's Resistance Army in Uganda and the Central African Republic. Likewise, the Haqqani network in Pakistan, Mexico's Zetas, and several Libyan militia groups can be considered conflict entrepreneurs.[12]

It is, of course, an over-simplification to imagine that FARC—a large, diverse organisation comprising multiple factions, fronts, and levels of engagement, from part-time members to professional guerrillas—is a monolithic bloc, and that all FARC leaders or cadres are motivated by personal gain. Many members of FARC (including numerous sympathisers and former guerrillas whom the authors interviewed for this project) hold apparently strong and genuine beliefs about social justice and political inclusion. Likewise, FARC leaders have committed decades to furthering the movement's aim of seizing power in Colombia, and their motivations are not simply plunder and profit. Yet, on balance—as the analysis presented in this book suggests—FARC's observable behaviour, from the field to the Secretariat level, suggests the organisation has become enmeshed in a complex war economy of its own making, affecting its goals and methods, giving it strong connections with organised crime, and creating what we might describe as a criminal-insurgent complex.

Defeat into victory?

By 1996, Colombia was losing the battle against this criminal–insurgent complex. Drug cartels—Pablo Escobar's organisation in Medellín, and the rival Cali Cartel—had subverted Colombia's democracy and brought violence to its cities. In the countryside, paramilitaries had merged into the Autodefensas Unidas de Colombia (United Self-Defence Forces of Colombia, AUC) and branched out into drugs, extortion and extrajudicial killing. In its modified version of Mao's Protracted People's War, FARC escalated from a stage of 'guerrilla war' to a 'war of movement', achieving a string of major victories between April 1996 and December 1999. FARC main force columns, operating openly in large formations, proved capable of defeating battalion-sized army units and seizing and holding territory.[13]

At the end of the twentieth century, by the army's admission, guerrillas controlled territory stretching 'from Ecuador to Venezuela, had built themselves considerable infrastructure in the southeast around Caquetá and Meta, and not only had Bogotá surrounded, but had deployed guerrillas into its outskirts. Road transport between the major cities was very difficult, if not impossible.'[14] FARC's victories and territorial expansion—which for the first time directly threatened Colombia's major cities—were a wake-up call for Colombians. Many had previously seen the guerrillas (to the extent they thought about them at all) as a nuisance; a problem for *campesinos* (peasants) but no threat to business-as-usual in Colombia's sophisticated centre. Suddenly the threat seemed real, prompting a national mobilisation.

Elected in 1998, President Andrés Pastrana initially pursued peace talks, creating a demilitarised zone—demilitarised, that is, for the government but not FARC—the size of Switzerland, centred on the area of San Vicente del Caguan, including a 'peace camp' at Las Pozos. But he broke off talks in February 2002, after the guerrillas had shown no willingness to abandon the armed struggle, had continued the fight outside the demilitarised zone, exploited the peace talks to gain breathing space, and used their Caguan enclave to massively expand cocaine production and attack Colombia's cities.

Colombian civilian and military leaders realised that something had to change, and began developing plans to break the deadlock. These eventually resulted in a major FARC defeat at Mitu, which signalled the government's new resolve and marked the beginning of Colombia's remarkable turnaround. President Pastrana had earlier, in 1998, formulated 'Plan Colombia' in partnership with US President Bill Clinton.[15] Initially conceived as a 'Marshall Plan for Colombia' and focused on achieving an end to the insurgency only insofar as this could be seen as part of a broader plan to reduce narcotics trafficking, Plan Colombia—formally known as the Colombia Strategic Development Initiative—'was a determining factor in the return of government control to wide areas of the country'.[16] Jointly Colombian and US-funded, and launched in 2000 with a budget (in year 2000 US dollars) of $7.5 billion, Plan Colombia was initially structured with 51 per cent of funding focused on institutional and social development, 32 per cent on counter-narcotics, 16 per cent on economic and social revitalisation, and roughly 1 per cent to support diplomatic efforts towards a peace settlement with FARC and others. Colombia initially committed to provide 65 per cent of the funds for Plan Colombia, with the United States and other international partners providing 35 per cent.

After the 11 September 2001 Al Qaeda terrorist attacks on New York City and the Pentagon, however, the focus of Plan Colombia shifted to expand cooperation beyond counter-narcotics into counter-terrorism. Encouraged by this boost, though largely relying on its own capability (Plan Colombia, though essential, nevertheless accounted for no more than 5 per cent of total Colombian government spending on efforts to counter the insurgency, deal with the illicit economy and extend government control), the Colombian security forces—military and police—began to turn the tide against the guerrillas, a process that hastened after the election of President Álvaro Uribe in August 2002.

Uribe took the fight to both guerrillas and paramilitaries, personally taking charge of the effort, turning the guerrillas' strategy—the 'combination of all forms of struggle' that treated armed action, agitation and propaganda, economic action, and political negotiation simply as facets of a unified struggle—against them through his concept of 'democratic security', implemented in his Política de Seguridad Democrática (Democratic Security Policy, DSP).[17] Under Uribe, and a series of talented and capable defence ministers, Colombia went from widespread insecurity to enjoying a greater degree of normality. Recruiting for the security forces surged: the armed forces grew from just under 205,000 in 2002 to 288,000 in 2013, and the police from 110,000 in 2002 to 178,000 in 2013. More importantly, the number of professional soldiers in the army (volunteers, as opposed to two-year conscripts) almost quadrupled from 22,000 in 2002 to 87,000 by 2010. The defence budget rose from 3 per cent of GDP to over 4 per cent during the 2000s, partly financed through a 1.3 per cent wealth tax on businesses and well-off Colombians.

Colombia's military improved in quality as well as quantity. New equipment—Black Hawk helicopters, Super Tucano attack aircraft, unmanned aerial vehicles, precision-guided weapons, and the latest communication and surveillance technology—paralleled the creation of a special operations command and increased investment in training. While the military and police bore the initial burden of conflict, follow-up efforts were led by the Centro de Coordinación de Acción Integral (Coordination Centre for Integral Action, CCAI), a reconstruction and stabilisation organisation supported by the US Agency for International Development (USAID) but embedded within, and reporting directly to, the presidency. As in earlier periods, low-profile civilian and military assistance from the United States helped—but the talent, energy and leadership that drove success were fundamentally Colombian.

Uribe tackled the criminal-insurgent complex—the nexus between the insurgency, paramilitaries and drug traffickers—through efforts to demobilise the AUC, which succeeded in 2006, prompting a dramatic drop in criminal violence. Much of this violence was also being perpetrated by FARC members, either operating on their own account or under the direction of insurgent leaders. The initial success of Democratic Security was seen in generating a huge drop in crime and violence from all sources, primarily achieved by the government's efforts to occupy and patrol all of Colombia's municipalities. By protecting communities that had previously seen little or no state presence, Uribe removed the main self-defence rationale for the paramilitaries. He also promoted demobilisation and reintegration of guerrillas, improvement of infrastructure, and popular dialogue throughout the countryside.

The effect was dramatic. Homicides halved from 28,837 (a rate of 70 per 100,000 people) in 2002 to 16,127 (35 per 100,000) in 2011; kidnappings plummeted by 90 per cent from 2,882 to 305; and car theft more than halved from 17,303 to 10,269 in the same time period.[18] The drop in kidnappings, in particular, brought a sense of relief and progress to Colombians. Security improvements helped the economy develop, creating a virtuous cycle of improving governance, economic growth and enhanced stability. Foreign direct investment rose to $19 billion by 2012, enabling further spending on security. Economic growth averaged 5 per cent annually during the ten years from 2002, enabling fresh investment in infrastructure, and funding for the expanding and professionalising military and police.

Uribe led a hands-on approach to popular dialogue, holding televised *consejos comunitarios* (community councils) each weekend across the country, where he and his entire cabinet travelled to small towns and city districts. This created a public forum that was both local and national, in which community members could directly pose questions to and raise concerns with the president and his ministers. The President also began to include in the national dialogue the marginalised communities that had been co-opted or intimidated by guerrillas. Uribe met with local communities in most of Colombia's thousand-plus municipalities and thirty-two departments in his eight years in office—visiting many areas more than once—and creating not just positive public relations, but a feedback loop that helped his administration fine-tune its policy through an active action-learning cycle.

Uribe's Democratic Security Policy was extended by his successor as President, Juan Manuel Santos—one of Uribe's last defence ministers—whose Sword of Honour campaign aimed to degrade FARC while consolidating control in 140 contested municipalities. Sword of Honour was developed in late 2011, using the principles of Operational Design, by a handpicked team that included some of the most gifted and battle-experienced officers in Colombia.[19] It called for increased pressure on FARC, quick-impact projects in contested districts (from water reticulation, sewers, bridges and roads, to community sports centres), and the initial creation of nine Joint Task Forces to take the fight to FARC bases. The later iterations of the plan (Sword of Honour II and III) included twelve Joint Task Forces intended to penetrate FARC strongholds, while territorial brigades and police secured contested areas, and civil agencies brought governance and development to normalising districts.

As inevitably happens in war, and as discussed in detail in later chapters, things have not worked out quite as planned. Still, the results are clear: there has been a steady increase in guerrilla surrenders and participation in the government's reintegration program, with 24,856 guerrillas demobilised between 2002 and 2014, and 1,350 guerrillas demobilised in 2013 alone. A further 6,000 FARC members have been killed in army raids, Joint Task Force deep-penetration operations, or precision strikes by the air force since the initiation of Sword of Honour. But the insurgents, under tremendous pressure, have not stood still. General Juan Pablo Rodriguez, chief of the armed forces in 2014–15, points out that 'FARC is not stupid. They adapt and change, and every day is more difficult for us.'[20] This can be seen in the most recent FARC numbers of approximately 6,920 full-time personnel and just under 10,000 part-time members as of February 2015, suggesting that despite all the pressure it was under, the organisation had still managed to recruit, replace losses, and continue to operate.[21]

Since 2010, the Colombian government has used operations Sword of Honour and Green Heart to steadily erode the strength and effectiveness of terrorist and criminal groups, targeting their violent methods and financial means. In terms of fighting structures, by the end of 2014 FARC had about 6,900 full-time fighters (a 25 per cent reduction since 2000); the ELN, 1,495; and BACRIM, 3,400—the smallest size of illegal armed groups in at least fifteen years. Fifty-five FARC leaders, seventeen leaders of the ELN, and forty-two from BACRIM have also been killed or captured through targeted operations in the same timeframe.

The reduction of violence and instability across the country is palpable. According to Colombian government data, 90 per cent of municipalities did not register any terrorist attack in 2014; 95 per cent experienced no subversive actions; 82 per cent of the population did not report presence of active terrorist structures or criminal gangs; and 94 per cent of the country had no recorded cases of kidnapping. As of 2014, only 6 per cent of the Colombian population was directly or indirectly affected by terrorist actions.

With respect to FARC's financial means and its materiel, the armed forces have also achieved notable success. Up until 2014, counter-drug operations had achieved a steady reduction of the area under coca cultivation to 48,000 hectares and the seizure of over half the potential cocaine produced from this crop. This, as well as the fall in kidnappings (and therefore extortion payments), represented strategic blows against FARC's funding. Regarding materiel, the armed forces seized 248 tons of explosives and 18,583 explosive devices (such as mines, mortar bombs and booby-traps), and defused or destroyed 69,411 improvised explosive devices over the period 2010–14. These losses helped the government roll back the territory and population controlled by FARC.

As it loses territorial control, FARC has been forced to drop back a stage in its strategy, abandoning the 'war of movement' approach adopted in the 1990s and returning to a combination of rural guerrilla operations and urban terrorism. Instead of frontal attacks on cities and military bases, FARC now hides among the population, using small local guerrilla groups or urban underground cells to snipe at soldiers, intimidate communities, extort businesses, and plant improvised explosive devices to deny access to base areas.

Against this backdrop, President Santos announced in his August 2010 inaugural address that the 'door to peace is not closed'. Peace talks, stalled since President Pastrana suspended them at the beginning of Plan Colombia, were restarted: exploratory talks between Colombian Government and FARC negotiators began in February 2012, and produced a six-point agenda for formal negotiations that began in Oslo, shifted to Havana in November 2012, and have continued into 2015. Santos made it clear that this time, unlike previous peace talks, military operations would continue until a deal was reached, and despite a unilateral FARC ceasefire (which the guerrillas themselves have, in fact, frequently breached) the government has continued its security operations throughout the talks to date. In addition to the peace process, the Colombian government has also made efforts to address the social basis of the conflict, through the 2011 Land Restitution and Victims' Law—which aims to redress human rights violations by all sides—and social programmes including Acción Social and a new Department of Social Prosperity.[22]

Relevance to counterinsurgency theory

Colombia can be considered a case of internal, or intra-state, insurgency and thus its relevance to counterinsurgency theory and to other countries' experience of conflict is worth explaining from the outset. Intra-state conflicts are geographically widespread and historically common. For example, the Correlates of War dataset, which tracks conflict since 1816, cites a total of 442 intra-state conflicts, many of which were insurgencies, over the past two centuries—more than for any other form of warfare,[23] Yet classical COIN theory—developed in the 1960s to address problems of colonial and post-colonial conflict under conditions of Cold War superpower confrontation—draws on a small and not particularly representative subset of these cases, while neo-classical COIN (revived after 2006 to help the United States and its allies cope with resistance warfare in Iraq and Afghanistan) also draws heavily on the same 'standard' campaigns—Malaya, Indochina, Algeria, Vietnam, the Philippines' Hukbalahap rebellion and the Mau-Mau campaign in Kenya.[24] Other cases are, of course, included in specialist and academic studies, and since 2006 the volume of writing on COIN in Afghanistan and Iraq has grown enormously, but to this day the official 'case theory' of COIN remains heavily skewed towards third-party campaigns, where First World nations send expeditionary forces to engage insurgents in distant, developing countries while also undertaking economic and political development of client governments.

This book is an attempt to break out of this conceptual straitjacket, to look instead at what happens when a government engages in extremely long-term operations, with only limited external support, against insurgents operating within its own territory. This set of circumstances creates a subtly but significantly different set of outcomes from those typically seen in expeditionary COIN campaigns, and is under-represented in the COIN literature. And yet, historically speaking, such campaigns (which, of course, include Colombia) may be more common—and thus their lessons more broadly applicable—than the more heavily studied 'classical' cases. In particular, given the authors' connections with Africa—and the opportunity, during fieldwork for this book, to bring several African civil and military leaders to observe and comment on conditions in Colombia—we feel that Colombia's experience holds significant lessons for African governments and military forces.

Parallels with African conflicts

The contemporary Colombia conflict started small but progressed within a generation to almost symmetric conventional battles, where the Colombian army was overrun by insurgent forces whose organisation and equipment was on a par with their own. In response, the Colombian state developed a 'whole of government' doctrine, strengthening and extending security and governance from the centre, aiming at building a more inclusive economic system in the process. Such an encompassing approach has been lacking, until now, in most African COIN efforts, for a variety of reasons—such as that these tasks are seen as a primarily policing or security ones, or that the security forces are aligned less to fighting than to rent-seeking from international resources, regional opportunities, and from their own fiscus.

But there are, too, distinct parallels. While the Colombian state never 'failed', by the admission of its own ministers it was at least on the cusp of failing in certain key respects. Indeed, Colombia's resonance with Africans and others springs from a number of characteristics that it shares with African conflicts.

First, Colombia (as noted) is not an expeditionary campaign in which European or American troops are engaged and drawing on relatively massive amounts of materiel; rather it is a campaign where local ownership of the problem and the response has been key to success. Second, like many African conflicts, Colombia's is a national struggle, with all the limits that this imposes in terms of sensitivity to civilian casualties, and, conversely, the advantages of local knowledge and time in which to solve strategic and political problems. Third, the Colombian example of extending governance and dealing with insecurity is topical and ongoing, not a historical discussion of events that occurred during the age of imperial decolonisation or against the backdrop of the Cold War. It's possible to talk to the players involved, to visit key locations that are still experiencing conflict, or to observe operations directly in the field, today. Fourth, the relationship between Colombia's under-governed (and sparsely populated) rural spaces, especially in the country's east, and the growth of revolutionary insurgent movements resonates with other countries where similar conditions exist—whether in Mali, Nigeria, Somalia or the Congo.

And fifth, Colombia exhibits all the symptoms of criminality mixed with insurgency. The changing relationship between criminality, guerrilla capability and government effectiveness has strong parallels in Africa: FARC and other guerrilla movements quickly became conflict entrepreneurs, with drugs as their major source of revenue, just as the same commercial metamorphosis occurred for Congolese and Central African Republic warlords, Al-Shabaab in East Africa and the Horn of Africa, Al Qaeda in the Maghreb, and Boko Haram across much of West Africa. At some point, in each of these conflicts, if not from the beginning, greed trumped grievance and self-interest outweighed ideology.

The focus of the book

With this as background—and even beyond its relevance to African conflicts and to other examples of 'home-turf counterinsurgency'—it's worth pointing out that Colombia is an understudied conflict, despite the contemporary fascination with COIN as a result of the 9/11 attacks and the wars that followed. Indeed, as the foregoing discussion suggests, the public perception of Colombia as a success story (specifically, an American success story) is significantly out of kilter with reality on the ground. This book sets out to describe the origins of the conflict and its various phases—with a focus on the period since 1999. It examines the evolution of guerrilla and counter-guerrilla strategy and tactics; highlights specific economic, political and military lessons; and identifies challenges in the transition from conflict to peace. It poses a question as relevant to Colombians as to anyone fighting insurgency and terrorism: What does victory look like?

If this volume shows anything, it is that Colombia is entering a new phase of struggle, a political war in which fresh challenges will emerge. Should a peace settlement be achieved, there is every likelihood that FARC will continue its efforts to seize power under another guise, perhaps applying a combination of subversion and open political agitation along with episodic or criminal violence (a strategy that aligns perfectly with the 'combination of all forms of struggle' discussed in Chapter 4 in more detail). Under these circumstances, the application of military force by the government will diminish in importance, effectiveness and relevance: it is impossible (not to mention immoral and illegal) to use lethal force against unarmed protestors or political cadres who operate in civilian clothes in the cities, manipulate social move-ments and apply primarily non-violent means.

As it enters this new phase of the struggle, Colombia will need new approaches that respect human rights, but which will still be effective to pre-vent the kind of takeover seen in Venezuela, Bolivia and Nepal by conflict entrepreneurs who simply adapt to new conditions. Ultimately, a more inclu-sive society, an economy that helps marginalised and excluded populations share in political and economic opportunity, the extension of rule of law and civil governance to every level of society, and economic policies that bring money, jobs and talent back to the areas that have suffered most heavily, are critical steps in conflict transformation. This will be extremely difficult—per-haps even harder than the military struggle—yet it will be utterly essential if Colombians are to achieve peace with victory.

1

A LONG WAR

David Spencer

This chapter offers a brief historical account of the main events, from 1949 to the present, in Colombia's long war against FARC. For reasons of space, it will not cover other groups (such the ELN, paramilitaries or M-19) in detail. The purpose of this chapter is not to substitute for other, more detailed, historical accounts available elsewhere, but merely to offer a historical frame of reference that explains the context for the conflict and sets out some of the campaign's key events, which we then analyse in more detail in subsequent chapters.

In this context, much depends on which date is chosen as the start of the conflict. Many US scholars take the years 1999–2000, when Plan Colombia began, as their starting point.

This is appropriate if the goal is to track that plan's progress, evaluate the effectiveness of US aid to Colombia, or claim that Colombia's success was in fact due to US assistance. But reality is a lot more complicated than this convenient (perhaps even self-serving) narrative suggests.

An alternative start point is 1982,[1] which is a sensible choice because several things converged in that year which still have an impact today: FARC developed its master strategy to take power, made the decision to begin taxing narcotics trafficking in the areas it dominated, and decided to participate in the first peace talks with the government of President Belisario Betancur, while simultaneously pursuing its plan to take power. Also in that year, drug-trafficking organisations began to process and transport large quantities of cocaine from Colombia to the US—and in response the government of President Ronald Reagan began to provide aid to Colombian authorities to combat the flow of cocaine into the United States. So, much current policy has its origins in 1982, even though the conflict is much older.

Other scholars start in 1966, when the 'Southern Bloc' guerrilla groups officially adopted the name Fuerzas Armadas Revolucionarias de Colombia (FARC). However, although FARC named itself at its second conference, its first conference was in 1964, after the Communist Party militias had been driven out of the rural enclave of Marquetalía. By that reckoning, the conflict is over five decades old—but if one includes the founding of the Communist Party's self-defence forces in El Davis and Villarrica, Tolima Department, this adds another ten or eleven years, making it six decades old. One could also go back to 9 April 1948, when Jorge Eliecer Gaitán was murdered in Bogotá, sparking the Bogotázo (riots in the capital, and the beginning of the civil bloodletting that became known as La Violencia). From there one could continue to go back further to the first Communist Party self-defence organisations formed in Viota in around 1930, making the conflict over eight decades old. This exercise could be continued almost indefinitely all the way back to independence or even the Spanish Conquest—maybe even further.

The point is that, in Colombia, there is almost always some important antecedent that has a significant impact on the next conflict or phase of the conflict, for this is a country that has continually been at war with itself.

Types of conflict and constraints

There have been many different types of conflict in Colombia, but (as discussed in more detail in Chapter 3) these have been subject to a number of common constraints, or constants. First, the Colombian central government has always been relatively weak. In fact, often the conflict has not been between the central government and non-government challengers, but between two or more non-government rivals. For many years the Liberal–Conservative wars were not included in the military history of Colombia because they had been fought not between standing, statutory armies, but between political militias. During La Violencia there was generally peace in the big cities where the government had significant military and police presence. However, in the countryside the bloodletting between Liberal and Conservative bands was ferocious. In the 1950s when the military sought to mediate between the groups and demobilise them, an observer remarked that 'Colombia was a strange country where the civilians made war and the military made peace'. The paramilitary–guerrilla war of the 1980s and 1990s echoes this phenomenon of weak central government and warring civilian militias. Thus the ability of Colombian governments and military forces in the twenty-first century to dominate and demobilise the modern civilian militias is a good sign, and perhaps signals the beginning of a shift in historical patterns.

Second, Colombia is rich in natural resources—gold, emeralds, coffee, cotton, sugar, bananas, oil, marijuana, cocaine and rare earth minerals—and most of its wars have revolved around some or all of these resources. Behind all the political banners and ideology has been the issue of the control and distribution of Colombia's vast natural wealth. Economics and politics are inextricably linked, as we explain in more detail in Chapter 5. If economy is defined as the distribution of goods and services, politics is the argument over the rules for the distribution of goods and services, and war or conflict is extreme politics. Conflict has always been intense in Colombia because the prize for the winner is vast. Furthermore, because of this vast wealth, there have always been abundant resources to purchase the tools of war or the knowledge and skills necessary to prosecute the war in the most clever and savage ways.

Third—as discussed in Chapter 3—Colombia's terrain is vast and fragmented, the size of France and Spain combined. Unlike many other Latin American countries, the broken terrain encouraged the development of regionalism. For many years Colombians identified themselves by what part of Colombia they came from: Cali, Medellin, Barranquilla, and so forth. The nation's fourteen major centres essentially developed in isolation from the rest of the country. The centre of most Colombians' world was not the capital, but their regional capital. Each region has its own customs, its own cuisine, and even its own accent. In other parts of South America people generally tell jokes about the idiosyncrasies of the inhabitants of neighbouring countries, but in Colombia these jokes are about neighbouring regions; Paisas tell jokes about Costeños, Costeños joke about Rolos, Rolos joke about Vallunos, while Vallunos—and everyone else—poke fun at Pastusos.

Within the regions themselves, the terrain is also vast and broken. This makes it very easy for a challenger to establish an enclave of power, and very difficult for national incumbents to winkle challengers out from their enclaves.

It is relatively easy for FARC to dominate those parts of the country where the central government is hamstrung by logistical and communications challenges in trying to expel FARC. Tactically, such expulsion is possible: it is relatively simple for the military to put together a force that can quickly capture terrain and population formerly under the control of the guerrillas. However, once this is done, sustaining government presence over the long term is nearly impossible. This is why the same territory has been fought over repeatedly and why areas cleared of guerrillas slip back into their hands a few months to a few years later.

The Culture of 'the Centre'

These difficulties experienced by Colombia's central government are exacerbated by a culture of 'the centre': national power is concentrated in the regional capitals or in Bogotá. For example, Barrancabermeja is Colombia's centre for oil production, yet very little of the oil wealth in Colombia is reinvested in Barrancabermeja—a miserable city fought over by militias and criminal gangs. Contrast this to Houston, Texas, a major oil hub in the US. People want to live in Houston because oil wealth is invested in the local community rather than taken away to somewhere else, whereas only those who have no choice stay in Barrancabermeja. The concentration of the nation's wealth into a few regional or national centres means that large numbers of people across Colombia are excluded from many of the benefits of the economy. This makes them ripe for recruitment by armed groups such as FARC and the ELN, as well as the paramilitaries in their day, and now criminal gangs. The military can go in and very easily retake these remote areas, but what these regions need in the long term is governance, not just military presence.

To give just one example: the best and brightest government bureaucrats are not sent to run the development programme in Santander de Quilichao, Cauca Department, a place that has seen much combat over the last three decades; such an appointment is regarded as a career killer. The most promising posts in Colombia's development programmes are not in the field, but with the Director of Social Action in Bogotá, or in the regional headquarters in Cali. No incentives, such as additional pay or guaranteed promotion, are offered to attract volunteers for the Santander de Quilichao post, or any of Colombia's other outlying towns. In fact, if there is such a post, the holder probably does not even reside there, but rather back in Cali, visiting the post once a month. Since nobody in the centre cares what happens in Santander de Quilichao, they are satisfied as long as things are not terrible. And so the position gets passed from government official to government official like a hot potato.

Yet small communities like Santander de Quilichao across Colombia are where the problem lies, and it is in these communities where the issues that drive the conflict will be resolved. Perhaps none of these local issues alone is enough to create a national problem, but their accumulation has a national impact, and develops into the violence that becomes a national problem.

In 2014, during a briefing, at a Colombian army divisional headquarters on the progress of the Sword of Honour campaign, the briefer stated that when the campaign began in January 2012, the FARC bloc in the area in question was known to have approximately 1,100 fighters. He also stated that his own headquarters' operational data showed the division had killed or captured over 900 guerrillas in the course of almost two years. These figures are not estimates. The Colombian security forces gather and analyse data meticulously. In other words, the 900 guerrillas in question were unquestionably real: either dead bodies or warm bodies in jails or demobilization programs. However, he went on to say that the FARC force they currently faced contained 980 guerrillas. In other words, the net reduction in guerrilla strength over the two years of the campaign was only 11 per cent. FARC had lost 82 per cent of its initial strength and yet had been able to replace 87 per cent of its losses. These figures demonstrate that, despite the dangers, incentives to join FARC are still strong among young men and women from rural communities. The lack of governance in these areas is a key reason for this. The government offers these people less in terms of security and economic opportunities than do the insurgents or violent criminal organisations.

This adds considerable tension to the military–civilian relationship on the ground. The military does not like continually conducting offensive actions against the same armed groups in the same areas, since this inevitably costs them dead and wounded. Furthermore, the enemy continually learns and develops more cunning and lethal traps for the advancing troops, making each repeated operation more costly for the army. The presence of the troops does not replace the need for good governance; it only opens the door for the civilian agencies. However, the civilian agencies are loath to join the military in rural areas.

Thus in the Sword of Honour campaign, military forces have reluctantly resorted to carrying out governance activities of their own. However, this is not their primary role and, though such activities are well intentioned and even well designed, the results are sub-optimal, a stop-gap measure at best.

The military and police on the front lines need to be accompanied by civilian agencies who can develop real solutions to the problems of governance in the rural areas.

The current war with FARC, the ELN and paramilitaries is drawing to a close. However, the next cycle with the criminal bands known as BACRIM, and FARC 'dissidents ('so-called 'FARCRIM'), is already well advanced. FARC, the ELN and even the paramilitaries claimed to fight for political causes. The new generation of armed groups may not make such claims; but, when examined in detail, the same causes that drove people to join the guerrillas will be the underlying drivers that cause people to join BACRIM and FARCRIM. Without local governance solutions—despite peace accords or other national solutions—the cycle of violence will continue.

The history of the conflict with FARC should be examined in this context. In its research, the historical commission appointed as part of the peace process divides the history of FARC into four phases between 1949 and 2013,[2] though for good reason this chapter divides FARC's evolution into six phases from 1949.

Phase one, from 1949–1966, stretches from when the Communist Party started recruiting militias during La Violencia to the official establishment of FARC; Phase two, from 1966–1982, sees FARC begin to evolve and progress from survival mode to beginning to develop a national strategy; Phase three, from 1982–1993, covers the consolidation and operationalisation of this strategy; Phase four, from 1994–2002, covers FARC's ascent and leap into mobile warfare; Phase five covers its descent and roll back under President Uribe's Democratic Security Policy from 2002–2008; Phase six, from 2008 to the present, covers FARC's adaptation in the face of central government successes.

Phase one: 1949–66

The first phase of the conflict saw the emergence of communist self-defence militias in the aftermath of the death of Jorge Eliecer Gaitán, their evolution, and the eventual formation of FARC after the fall of Marquetalía.

The bipartisan violence unleashed throughout the country by the murder of Gaitán on 9 April 1948 led to the formation of Liberal guerrilla groups, generally in resistance to local Conservative Party bosses and their militias. In Tolima, in the mountainous region between Bogotá and Cali, the Communist Party began to recruit some of these guerrillas by offering them political-military education and training. Among them was Manuel Marulanda, the eventual first secretary general of FARC. However, communist influence was not to the liking of the Liberal Party, and this led to an internal conflict within the Liberal guerrillas. Those that were against the communists helped the army and Conservatives attack them at their base community in El Davis. From El Davis, the communist guerrillas moved to Villarrica, Tolima.

Then began the great Liberal guerrilla demobilisations under the military government of General Gustavo Rojas Pinilla (1953–57)—the only military government in Colombian history. While the Liberal guerrillas demobilised in large numbers, the communists refused to do so. In response, the government launched a large-scale offensive, expelling them from Villarrica. The survivors were displaced to upper Sumapaz, El Pato, Guayabero and Marquetalía, where they established new communist settlements.

There was a lull for a few years, until in 1962 the Colombian Army's 6th Brigade began unsuccessful operations against the guerrillas in Marquetalía. In 1964 the government declared that there were 'independent republics' in Colombia that could not be tolerated and launched Plan Lazo, a counter-guerrilla campaign designed with the assistance of US advisers who were veterans of the Hukbalahap Rebellion in the Philippines.[3] After expelling the communists from Marquetalía, the government continued its operations against the communities at Rio Chiquito, El Pato and Guayabero. The surviving guerrillas of these communities got together and held their First Conference at the end of 1964, calling themselves the 'Southern Bloc'. They declared that their objective was to seize national power via guerrilla warfare; they also adopted Law 001, which called for comprehensive land reform. In 1966, guerrillas of the Southern Bloc held their Second Conference and were thereafter known as 'FARC'. To achieve their goal of fomenting a nationwide revolution, they sent cadres to different parts of the country to set up guerrilla columns.[4]

Phase two: 1966–82

During the second phase, FARC was essentially in survival mode, attempting to figure out how to fight a guerrilla war and gather the necessary resources to survive. This phase concluded just before FARC's Seventh Conference.

Between 1966 and 1982, FARC raised money mainly through extortion and kidnapping, gradually developing experience and new plans. Militarily, it conducted low-level guerrilla attacks and ambushes. Most of FARC's advances were incremental, as reflected in the conclusions of its periodic conferences. During the Third Conference in 1968, FARC established a National Training School for cadres and began discussing lessons learned from other conflicts, especially Vietnam. During the Fourth Conference at the beginning of 1970, FARC leaders made additional adjustments to the organisation—promoting, redistributing and expanding various individuals and forces. During the Fifth Conference in 1974, they began to think about the contribution that organised masses could make to the guerrilla movement. Perhaps most importantly, they began to develop clear guidelines on the nature of guerrilla fronts, which became the basic underpinning of FARC's organisation—although they had still not developed clear guidelines on general staff structures, nor formed a superior national-level command.

The Sixth Conference, held in 1978, was the most important conference of this period and probably one of the most important in FARC's history, though it is relatively unknown because it was eclipsed by the Seventh and Eighth Conferences. FARC's statutes were approved at this conference, as were the regulations for front staff and the national governing body, the Central High Command. An emerging plan for seizing power was approved that was later consolidated at the Seventh Conference. Some sources indicate that the initial relationship with drug trafficking began as early as 1980, when drug traffickers offered to pay FARC to provide security to their drug processing laboratories.[5] Over these sixteen years FARC grew slowly to 1,200 combatants organised into sixteen fronts.[6]

Phase three: 1982–93

In 1982, at its Seventh Conference, FARC developed a strategic plan to take power. At the Eighth Conference (in 1993), FARC leaders operationalised this strategy. Between the seventh and eighth conferences they attempted to build their mass base by participating in the first peace process with the Belisario Betancur government in order to develop a national audience.

When representatives of what would become the Betancur administration queried FARC as to their interest in peace negotiations in 1982, FARC leaders were in the midst of planning their strategy to take power. Influenced by events in Central America, FARC believed that the international situation was such that a revolutionary process could be stimulated in Colombia. But for this to be successful, the revolutionary organisation, with FARC at the vanguard, would need a strategic plan. FARC had to build two things in order to make the jump from a guerrilla to a mobile war: a front-type revolutionary political party and a revolutionary army. FARC did not see any contradiction between dialogue and negotiations with the government while simultaneously developing a strategy for taking power.[7]

FARC adhered to a concept of combining all forms of struggle so that the method to take power was not exclusively military. Rather, different combinations of military, political, economic and social means could be used to get the guerrillas closer to their objective. Due to Colombia's bloody political history, FARC believed that the only way to ultimately achieve power was through violence, but such an offensive could be supplemented by political struggle. If violence was not FARC's main form of struggle, it would be at a minimum their contingency plan. Thus for FARC, the peace process was a means of significantly strengthening the political struggle. First, it would allow the organisation to share its political message openly with all Colombians as well as the global community. Second, participation in elections and national debates would facilitate military efforts.

Initially, FARC was very successful in the political arena. They created the Unión Patriótica party (Patriotic Union, UP), established branches all over the country, and in the elections of 1986 won seats in the Senate and Chamber of Representatives, as well as coming in third in the presidential election.[8] In the interest of inclusion, the government of President Virgilio Barco appointed UP mayors in all of the areas where FARC predominated, and even in some places where the Liberal Party had won majorities.[9]

However, FARC's understanding of the UP was very different from the expectations of the rest of society. Where FARC saw the UP as complementary to military action, government and society expected that UP political activity would replace FARC military action. When it did not, some people both inside and outside of the government became suspicious of FARC treachery and began to carry out irregular activities of their own; in parts of Colombia, UP members were harassed, threatened and attacked. Yet this only served to reinforce FARC fears that the peace process was a plot to defeat the movement by putting it in a position of helplessness, as well as encouraging the perception of FARC leaders that military action continued to be necessary. FARC's intransigence towards demobilisation only provoked more intense attacks and violence against the UP, and so the unhappy downward spiral continued.

Militarily, FARC decided to formally move from guerrilla warfare to mobile warfare. To do this, it needed to make major organisational changes as well as develop a military campaign plan. In 1982, the organisational change was marked through the addition of the phrase 'Ejercito del Pueblo' (People's Army, EP) to FARC's name, so that it became FARC-EP. The organisation also adopted a new tactical doctrine it called the 'Nueva Forma de Operar' (New Mode of Operations, NFO).[10] At this time, FARC-EP numbered 1,200 combatants, and its leaders realised that in order to become an army they needed to greatly expand their forces. To do this they needed a recruiting pool and a source of financial resources. Both of these needs were fulfilled by a new phenomenon in FARC-dominated areas: drug trafficking.

At the start of this period, coca was predominantly grown in Peru and Bolivia and was flown into Colombia for processing into cocaine. The vast and largely uninhabited eastern savannahs and southern jungles were ideal because coca processing could prosper without too much interference from the authorities. These regions also represented a convenient bridge location between producing countries (Bolivia and Peru) and consuming countries (mostly the US and Europe). The proximity of jungle meant that there were plenty of places to hide labs, and the savannahs were ideal for building air-strips—where raw materials could be brought in and refined product could be shipped out.

However, this was FARC-controlled territory, where the organisation had been forced to move as a result of Plan Lazo. Here FARC had scraped together a living by extorting cattle ranchers, but the influx of drug traffickers along with poor Colombians in search of work greatly altered their prospects. FARC identified the poor migrants as a source of recruits and the drug traf-ficking as a source of income. Ultimately, FARC leaders wanted to get money from countries within the international socialist bloc—as Central American groups had—but until then they had to make do with what was available. They decided to impose taxes on the 'evil capitalists', the designation they gave to the drug traffickers.[11] This was supposed to be only a temporary measure until foreign socialist countries could be convinced to provide aid. Little did they realise the impact that drug trafficking would eventually have on their own organisation.

FARC had ambitious plans for its army. After many years of secrecy, the campaign plan (which was an operational-level plan, subordinate to the stra-tegic plan developed at the Seventh Conference) was finally committed to paper and shared with the rest of the organisation in 1989. This plan has been referred to in FARC documents by many different names: as the 'Campaign Plan', 'Eight Year Plan', and finally in 1989 as the 'Bolivarian Campaign for the New Colombia'.

This plan was supposed to take place in four phases, each lasting two years, as follows.

During the first phase, FARC was to grow to sixty fronts of 300 combatants each until it had a force of 18,000. During the second phase, FARC was to increase its numbers to eighty fronts of 400 combatants each, for a total of 32,000. During both of these phases they were to seize control of the Eastern range of Colombia's three north-south mountain ranges (the Western, Central, and Eastern Cordillera). The Eastern Cordillera was to be their 'Centre of Strategic Deployment' and to be used as a corridor to move forces to their strategic objective, Bogotá.

During the third phase, FARC would launch their first strategic offensive, sending 16,000 guerrillas north across the Eastern Cordillera to attack the capital. The other 16,000 guerrillas would attack targets across Colombia to tie down government forces and prevent them from reinforcing Bogotá.

The fourth phase was a back-up phase. Anticipating that the first general offensive might not succeed, FARC columns would withdraw into their base areas and, while on the march, organise a second strategic counter-offensive, which would presumably prove victorious.[12]

When FARC created this plan in 1982, its troops numbered 1,200 guerrillas organised into twenty-seven fronts. During the peace process, FARC leaders moved aggressively to prepare for the plan's operationalisation, so that over seven years (at the end of which they made the plan known internally) they grew to forty-four fronts, essentially doubling their forces. FARC leaders optimistically reported that they were close to completing the first phase (an essentially preparatory phase) of their campaign plan.[13] As they grew, FARC columns also began to systematically expand their territorial control, particularly in the eastern plains. In these regions, guerrillas who had once been seen as interlopers began to aggressively impose their will by harassing, extorting, exiling, kidnapping and murdering elements they considered undesirable or troublesome.[14] It was also at this time that the relationship between FARC and drug traffickers was consolidated. Their initial encounters were resolved relatively peacefully, and the drug traffickers began paying FARC not only to leave them alone but also to provide security to their labs and landing strips, thus providing a significant boost to FARC finances.

Yet it was also during this expansion in the 1980s that the father of a rancher called Carlos Castaño was kidnapped and murdered by FARC in Antioquia, spurring Castaño to found his own anti-FARC paramilitary group—he would eventually become the nominal national leader of all of Colombia's paramilitaries.

Similarly, the relationship between FARC and the drug trade was not straightforward. FARC wanted more money from drug traffickers and in 1987 they stole cocaine from Gonzalo Rodriguez Gacha, a narco in the southeastern plains of Colombia. Gacha sent a warning to local FARC guerrillas to return the cocaine; when they did not, Gacha retaliated by carrying out an assassination campaign against the UP—FARC's most vulnerable element. Gacha was assisted by individual members of the security forces, as well as the paramilitaries under his influence and control. More than 1,000 murders of UP members were officially registered, though human rights advocates claim the real total was over 3,000.[15] The degree to which these killings were carried out by drug traffickers, paramilitaries, or rogue members of the state is still unclear. President Virgilio Barco's government did offer the UP politicians protection from the Departamento Administrativo de Seguridad (the domestic security force, Colombia's equivalent of the FBI) and police bodyguards, but in some cases these were refused outright and in others the UP politicians were killed when they failed to take their security details with them.[16] The combination of murders and popular disillusionment with FARC resulted in the dissolution of the UP.

In August 1990, shortly after the election and inauguration of César Gaviria as President of Colombia, Jacobo Arenas, FARC's political theorist and co-leader (with Manuel Marulanda), died. FARC's political thinking would essentially remain frozen with the death of Arenas. The end of the Cold War also had a major impact on FARC leaders, because they lost hope of receiving assistance from the international socialist bloc, and began to rely increasingly on drug trafficking. Between 1990 and 1991, FARC conducted a series of increasingly violent attacks as part of the Coordinadora Guerrillera Simón Bolívar (Simón Bolívar Guerrilla Coordinating Board, CGSB) offensive, a grouping of FARC, EPL and ELN guerrillas. The most successful of these was the attack on the military base in Taraza in November 1990. This provoked President Gaviria to take a harder line towards FARC, and in December 1990 he ordered the military to take Casa Verde—a quasi-demilitarised zone (DMZ) that had been granted by the government of President Betancur for the peace negotiations. However, despite fierce fighting, the army was not able to kill or capture any of the FARC leadership.

In response, the CGSB launched a series of offensives in the first half of 1991. While the vast majority of the actions were guerrilla and terrorist acts, there were a few differences between the CGSB campaign and previous guerrilla offensives. First, the new offensives were considerably larger in scale. The

CGSB guerrillas were able to launch two-dozen simultaneous attacks across multiple fronts.[17] Second, they were able to sustain this effort for several months. Attacks against infrastructure increased, particularly the electrical grid and oil pipelines, and kidnapping became a common tactic. A few more spectacular attacks were attempted, such as the successful taking of the Girasoles military base in March 1991. The increased intensity of the fighting was evident: in 1990, for example, the guerrillas attacked the government 169 times, killing 339 members of the security forces. In 1991, the guerrillas were able to carry out 425 attacks, killing 461 and capturing 276 security force personnel.[18]

In the midst of this offensive, the government attempted to continue its policy of seeking peace negotiations. Representatives of the CGSB met with the government in Cravo Norte; then in Caracas, Venezuela; and later in Tlaxcala, Mexico. The government was pushing to turn the talks into demobilization accords, but the guerrillas refused to renounce the armed struggle. They wanted to focus on international humanitarian law and the 'humanisation' of the conflict. With no agreement, the dialogue ended when the EPL kidnapped a former government minister who would later die in captivity.[19] The demise of the peace talks also signalled the demise of the CGSB, with its constituent guerrilla organisations gradually going their separate ways.

In late March and early April of 1993 FARC held its Eighth Conference, which focused on operationalisation of the strategic plan developed at the Seventh Conference. The FARC leadership now ordered the operational execution of their strategy, and of the campaign plan developed alongside it. The new plan was called 'Jacobo Arenas; We Are Fulfilling!' in honour of the deceased FARC ideologue.[20] In addition to ordering the execution of their strategy to take power, FARC cadres developed an alternative (or replacement) government platform, which they called 'A New Government of Reconciliation and National Reconstruction'.

Analysing the fall of the Berlin Wall, FARC vehemently defended the armed struggle, accusing President Gorbachev of treachery and claiming that what was going on in Europe did not apply in Colombia because deep social problems still existed that could not be solved peacefully.[21] The conference also officially abandoned the UP and the official (legal) Colombian Communist Party (PCC). Further political efforts would be made through the development of a clandestine political apparatus, Partido Comunista Clandestino Colombiano (Clandestine Communist Party of Colombia, PC3), discussed in more detail in Chapter 3.[22]

FARC ordered its now sixty fronts to recruit enough people to reach the planned strength of 32,000 guerrillas and to bolster the urban networks. At the time, FARC's urban networks were probably among the least developed of it units, but they were vital to its strategy for taking power, which was meant to climax in urban offensives.[23] In addition, the fronts were ordered to build militias in their base areas to protect them from recovery by the government. These militia units were also very weak.[24]

Developing international relations, particularly to gain material support, had been one of the priorities of the strategic plan. The Eighth Conference created FARC's international commission (COMINTER). FARC had experienced many difficulties in gaining international support and had been able to formalise relations with only three countries: Cuba, Nicaragua and Vietnam. Cuba provided limited military training and agreed to treat FARC casualties, Nicaragua allowed FARC to use its territory for administrative purposes, and the Vietnamese had given conferences and training classes to FARC to compare experiences and teach FARC cadres about communist victory in Vietnam. FARC was especially appreciative of the Vietnamese assistance.[25] However, it had not been able to acquire material support from any country.

In addition to relations with foreign governments, FARC also developed relationships with revolutionary parties and movements. FARC cadres gave military training to Venezuelans, Ecuadoreans, Brazilians and Peruvians. The Chilean Communist Party sent members of the Manuel Rodriguez Patriotic Front (FPMR) to Colombia to provide training, but FARC leaders found them intolerably arrogant. FARC leaders ordered international work to intensify. The organisation had carried out financial operations in Venezuela and Panama and conducted military operations in Brazil, Ecuador and Peru—though later, in retrospect, FARC leaders were to consider these cross-border military actions a mistake.[26]

Militarily, FARC emphasised several points at its Eighth Conference. First, it planned to carry out bloc-level operations. This meant carrying out multiple, simultaneous tactical operations at the regional level. To do this it created a total of five blocs: the Jose Maria Cordoba Bloc (also known as the North-Eastern Bloc), the Caribbean Bloc, the Middle Magdalena Bloc, the Eastern Bloc, and the Southern Bloc. Two additional organisations similar to blocs were also created: the Western Joint Command and the Central Joint Command. Whereas blocs contained more than five fronts, the joint commands contained fewer than five—though it was foreseen that the joint commands would eventually become blocs. (The similarity has caused many

analysts to refer to FARC's 'seven blocs', which for most purposes is correct.)
The Eastern Bloc would make the main effort in the direction of Bogotá,
while the Southern Bloc would be a primary source of finance, and the rest of
the blocs would carry out attacks to spread and dilute the government forces
(an approach some FARC leaders called the 'centrifugal strategy'). To make
sure that the blocs were run correctly and carried out the strategic plan, a
member of the FARC secretariat was to be assigned to lead each one.[27]

The creation and operation of FARC's army required significant resources.
In 1989, FARC contemplated that each phase of the Eight Year Plan would
cost $52 million, over and above existing revenue, for a total of $208 million.
Acquiring this additional money was a major concern for the FARC leadership. Fronts were encouraged to obtain more resources from the oil, mining
and electrical power companies. Essentially, they were to do this through
extortion and kidnapping: the FARC secretariat ordered that 'All of the fronts
and blocs should examine carefully the different places where there is mining,
etc., so we can send units to begin controlling the area and in this way see how
we can begin to force them to pay taxes for our sustainment.' Fronts were also
told to generate their local economy by investing in local businesses, such as
gold mining and the cattle market. The leadership did not mention drug trafficking, but euphemistically referred to it as an unpopular method of financing since it delegitimised the organisation.[28]

FARC had been conflicted from the very beginning over the immorality of
the drug business, and constantly encouraged the fronts to attempt to lower
their dependence on narcotics. However, nothing else provided as much
money; and rather than divest themselves of the illicit business, FARC leaders'
dependence on this method of financing continually increased. FARC was
now on the verge of launching its strategic offensive.

Phase four: 1994–2002

The fourth phase saw the implementation, and eventual defeat, of FARC's
mobile warfare campaign—the so-called 'war of movement'.

While the government focused on combating the Cali Cartel, FARC began
to systematically increase the tempo and intensity of its operations. FARC
operationalised the conclusions of its Eighth Conference, with blocs carrying
out coordinated offensive operations. They systematically cleared mobility
corridors of government forces by launching repeated attacks against police
stations. The guerrillas did not always take the stations, but the damage they

caused forced over 150 to be abandoned. In this way, large swaths of the countryside were abandoned to guerrilla control or influence.

Politically, FARC implemented what they called 'co-government'. FARC local cadres often dictated local-government policy, spending, and political candidates in areas where they dominated, without ruling directly. What they failed to realise, however, was that this did not translate into real support, as most people complied out of either fear or for personal profit (in the form of bribes and kick-backs).

Although the military had noticed an increasing intensity of combat, it was truly shocked on 30 August 1996 when all seven FARC blocs launched twenty-six simultaneous attacks across the country. The most intense violence occurred in La Carpa and Las Delicias. This was the beginning of what the guerrillas called the 'war of movement'—the second phase in the Maoist Protracted People's War strategy, or Vietnamese Prolonged Popular War. While most of the action continued to be traditional guerrilla warfare—ambushes and assaults on police stations in small towns—the intensity was new. Actions took place almost daily in multiple locations across the country. These were punctuated by truly mobile-warfare actions like that at San Juanito in February 1997, and at Patascoy Hill in December of the same year.

In June 1997, having been unable to liberate the sixty prisoners captured at Las Delicias, the government made a deal with FARC to demilitarise Cartagena del Chaira and Remolinos del Caguan, and in turn received the sixty captives from Las Delicias and twenty more that FARC had captured around the same time in Antioquia. This event humiliated the military because it sent the message that they were incapable of fighting the guerrillas, and that progress could only be made through demilitarising special zones and negotiating.

The FARC 1997 plenum indicated that while it was taking the movement longer than the expected eight years to accomplish its objectives, the fronts were making good progress.[29] They had created 200 mobile companies of the projected 600 (approximately 10,000 combatants out of a planned 30,000), and were moving forward even though implementing the New Mode of Operations, or mobile warfare, was causing some trouble. FARC leaders now proposed the creation of mobile blocs—groupings of ten or more mobile companies—to act against strategic targets. The experiments with simultaneous offensives across the country were considered successful, and more were ordered. FARC established the objective of 'liberating' the departments of Caquetá and Putumayo to then set up a provisional government to receive international recognition and belligerent status.[30]

In 1998, the intensity of FARC offensives increased. In March they intro-duced a new and terrifying weapon—known to FARC as the *rampla*—which the press immediately dubbed the gas-cylinder bomb. The *rampla* was a crude homemade mortar originally designed by the Irish Republican Army (IRA) and manufactured from natural gas tanks. The IRA sent advisers to Colombia to teach FARC how to make and use them.[31] They were highly inaccurate, but when fired in numbers they could sweep away anything in their path, and could penetrate existing government fortifications.

FARC debuted the new weapon in March 1998, overrunning an under-strength battalion of the 3rd Mobile Brigade at El Billar. This event was more significant than the losses at Las Delicias, as the army's Mobile Brigades were supposedly elite forces.

On 4 August 1998, FARC launched its biggest offensive of the conflict, consisting of more than thirty attacks in seventeen departments. While most failed to overrun their targets, the offensive caused much damage, and suc-ceeded in overrunning the Miraflores counter-narcotics base in Guaviare Department, as well as an army base at Pavorando, Antioquia. Dozens of soldiers and policemen were taken hostage. Many FARC operations were aimed at capturing army and police for an eventual prisoner exchange. FARC pursued this tactic because it was one of the necessary steps to achieving bel-ligerent status, thus allowing them to establish formal relationships with for-eign governments.[32] In turn, this status would allow them to openly import weapons and equipment.

As President Andrés Pastrana took office, FARC continued to attack the military. Nevertheless, the government moved forward with establishing the El Caguan DMZ, which had been allocated to FARC as part of the peace process.

On 1 November 1998, days before the DMZ was to be inaugurated, FARC launched an offensive as a show of strength, attacking the department capital of Mitú, in Vaupes. Mitú was no more than a small town, but it was a depart-ment capital nonetheless. Located in extreme southeastern Colombia, it was surrounded by an area of drug trafficking, and, more importantly, the air force could not reach it without refuelling and no adequate intermediate air base was available. Initially, the attack went well for FARC, but what they did not count on was the air landing of a large relief force twenty-eight hours later, which retook Mitú and inflicted heavy losses on the guerrillas. The Colombians had negotiated with Brazil for the temporary use of a runway just across the border.[33] The defeat at Mitú proved to be the beginning of a series of setbacks for FARC, the most important occurring in July 1999 at Puerto Lleras and Puerto Rico, in Meta Department. Nevertheless, the guerrillas continued their attempts to improve their military position. They launched several additional offensives through 1999 and 2000, all of which involved significant attacks from the refuge of the DMZ. FARC made multiple attacks in November and December of 1999; in January, February, and March of 2000, it renewed attacks in the direction of Bogotá. The problem for FARC was that it repeatedly used the same tactics, so while the combat was intense, the army had already adapted and learned to defeat these methods.

In the midst of this offensive, FARC held a plenum of the High Command in March 2000. FARC leaders congratulated themselves on the qualitative leap they had made in fulfilling the Strategic Plan and their role in 'opening up new spaces for struggle for the blocs and fronts across national territory'.[34] The leadership toyed with the idea of a ceasefire with the government, but only if it allowed them to locate their forces closer to the major cities and achieve their goal of belligerent status.[35]

Politically, FARC created the Clandestine Communist Party of Colombia (PC3) and the more inclusive (and overt) Movimiento Bolivariano por la Nueva Colombia (Bolivarian Movement for a New Colombia, MBNC). The MBNC was to be composed of the PC3 and all those people who sympathised with FARC but were not necessarily part of the organisation or the party.[36] This movement was officially inaugurated with great fanfare in the DMZ on 29 April 2000. FARC leaders also ordered more work on the international political front. They would invest money in reviving many of the flagging regional communist parties with the hope of both revitalising the continental revolutionary movement and broadening space for FARC.[37]

At this time, also, FARC became more involved in narcotics—particularly coca growing, cocaine processing, and international distribution through alliances with drug-trafficking organisations. FARC controlled all aspects of the business. It did not allow unauthorised transactions of any type in its areas of influence and swiftly punished anyone who attempted to break the monopoly. FARC controlled the coca buyers, who bought the entire product from the farmers, and it charged taxes at all levels of the growth and production process. Cartagena del Chairá was one of FARC's main coca producing areas in 1999, with 6,405 detected hectares under coca cultivation. After the establishment of the DMZ, this area rose by 112 per cent to 13,551 hectares—surging from an estimated annual production of 37 to 78.5 tons per year of cocaine. It was the second-highest coca-producing municipality after the Guamuéz

Valley in Putumayo, which had 16,523 hectares under cultivation.[38] In 2000, the Integrated Illicit Crop Monitoring System (SIMCI)—a monitoring program run by the United Nations Office of Drugs and Crime—detected 21,759 hectares of coca in Putumayo, a growth of 28 per cent from the previous year, all controlled by FARC.[39] Laboratories to produce cocaine multiplied in the DMZ, as did illegal flights into and out of the zone. Most flew to the DMZ from Barrancominas in Guainía, Colombia's most easterly department, which had become one of the major hubs for the importation of precursor chemicals into Colombia, due to its proximity to Venezuela and Brazil.

As the FARC mobile warfare offensives failed to gain traction, the rebels turned increasingly to terrorism. In September 2000 they hijacked a small commuter airliner and, a month later, declared a transportation blockade in Putumayo Department in protest against the launch of Plan Colombia, and attacked multiple town centres with *ramplas*, displaying no regard for noncombatant casualties. In December 2000, FARC murdered Congressman Diego Turbay Cote and members of his family.

In 2001, the first undeniable evidence of FARC's link to international drug trafficking became public during Operation Gato Negro, carried out around Barrancominas, Guainía. During this operation, Fernandinho Beira Mar (a Brazilian drug trafficker) was captured by the military. He turned out to be the main point of contact with FARC for shipment of drugs through Suriname to international markets and for the importation of weapons to FARC. Because of the organisation's increasing links to international drug trafficking, the US then issued its first requests for extradition of FARC leaders.

In the same year, FARC unilaterally released several hundred enlisted soldiers it was holding as prisoners and shifted its kidnapping strategy to 'political' targets. Subsequently, FARC members kidnapped Alan Jara (ex-governor of Meta) and Consuelo Araújo (former Minister of Culture, whom they later murdered). The international connections of FARC also became evident when three members of the IRA were captured in Bogotá. They were released eventually on legal technicalities, but their capture highlighted the fact that the IRA was helping FARC learn techniques for urban terrorism, which FARC needed for the next phase of its struggle.[40]

FARC continued attempting to launch major mobile attacks from the DMZ. In June 2001 they carried out a successful attack at Coreguaje, Putumayo.[41] In August they attempted to march from the DMZ to recover Barrancominas, their former logistics and drug-trafficking hub—only to be soundly defeated by the army in Operation 7 de Agosto.[42] After this defeat of a sizeable force attempting large-scale mobile warfare, FARC reverted to guerrilla-warfare and terrorist (that is, urban-asymmetric) tactics.

In January 2002, in view of its recent losses, FARC held an extraordinary plenum. In it the organisation's leaders confirmed their objective in the peace talks—to 'take power by using the dialogue for the proposal and implementation of mechanisms to facilitate that goal.'[43] To force the issue, they ordered their troops to carry out sustained operations against businesses and the infrastructure of the oligarchy. This was to include the electrical grid—particularly those parts that supplied the big cities—and bomb attacks on army bases and police stations.[44] However, rather than forcing the government to cede ground, these moves hardened attitudes towards FARC. The straw that broke the camel's back came in February 2002, when four rebels hijacked a commercial flight between the town of Neiva and Bogotá and forced it to land on a remote country road, taking thirty passengers hostage including the Senator Jorge Eduardo Gechem Turbay.[45] This brazen attack prompted the government to break off peace talks the very same day—President Pastrana ordered the army to recover the DMZ. The military moved in almost immediately, and FARC withdrew from the zone. During the withdrawal, however, FARC kidnapped presidential candidate Ingrid Betancourt at a roadblock.

FARC's response to the termination of the peace process was twofold. On the one hand, it attempted to make governance difficult, if not impossible. The guerrillas notified mayors, councilmen and aldermen that if they exercised their office they would face kidnapping and death.[46] They also increased urban terrorism, focusing on the roads and infrastructure that connected to the rest of the country.[47] Perhaps their most brazen move occurred in Barranquilla in April 2002, when they tried to influence the presidential election by attacking candidate Álvaro Uribe with a bomb. This backfired spectacularly, however, when the attack failed and Uribe was subsequently elected as president.

Phase five: 2002–08

During the fifth phase of the war, and following the failure of its 'war of movement', FARC found itself being forced into a definitive military retreat.

FARC greeted President Uribe's inauguration with a new wave of terror, attempting to launch 180 homemade mortar-bombs at the ceremony. Most failed to launch, and of the approximately twenty that did most were misses, hitting the military academy building and a group of homeless people near the inaugural ceremony. Despite this aggressive beginning, FARC was now on the defensive.

Manuel Marulanda sent an angry letter to FARC cadres, urging the guerrillas to confront Uribe. Morale was down and mistakes had been made, so he chided his commanders about having forgotten the fundamentals of irregular warfare. He scolded them for not understanding the strategic plan, for having underestimated the enemy, and for having gone soft. Marulanda dismissed President Uribe's new policies and ordered his commanders to wage a war of attrition through terror, guerrilla warfare, and the massive use of mines. They would control local politics by killing or kidnapping mayors and councilmen who dared do business against FARC's will.[48] As FARC secretariat member and military commander Mono Jojoy put it in a subsequent document, 'Increase carrying out the order against the institutions, whoever does not resign, give him lead [*pistola*] but do something so they believe us.'[49]

However, FARC's leadership failed to understand the fundamental difference between the approach of President Uribe's government and previous government operations. The new government strategy was not just an increase in troops or an increase in operations, but a fundamentally different way of thinking about security. Where previous governments had acted in a piecemeal fashion against guerrilla organisations, Uribe's Democratic Security Policy approached Colombia's security problems with a truly whole-of-government effort—not delegated to the military, but led by the president. This new approach began to have a significant impact on FARC.

In February 2003, FARC tried to derail the DSP by bombing the upmarket social club El Nogal in Bogotá. The bomb killed thirty-six people and wounded more than 200. The purpose was to make 'the rich feel the bombs in their own houses',[50] according to Mono Jojoy. But this was a spectacular miscalculation: Colombia's elites did indeed feel the effect of the insurgency, but when they felt it, the bombing galvanised them and consolidated their support for Uribe's policy against FARC.

However, despite the initial setbacks, FARC's ability to project strategic force from its base areas (which, for the most part, remained undisturbed by military operations) was essentially intact. The military soon moved against these base areas, first around Bogotá and then in the area between the Meta and Caqueta departments. This plan was executed in two phases. The first, Operation Libertad I, began in May 2003. By October, Aurelio Buendia, the FARC area commander, had been cornered and killed. In FARC's own after-action review, the organisation admitted that it had been defeated by the new methods.[51]

In March 2004, Operation JM, the second phase of the offensive, was launched against the area between Caquetá and Meta known as 'the

Secretariat'.[52] FARC developed a counter-plan known as 'Resistance Operation Urias Rondon'. It was essentially an attrition operation that used snipers, mines and ambushes to deny military access to the base areas. However, this plan failed and the guerrillas suffered heavy losses. After the first six months, FARC commanders held a conference in which they admitted that more than 500 of their guerrillas had been killed, captured, or had deserted, and that they had lost a large amount of materiel.[53]

In 2004, the government formed Joint Task Force (JTF) Omega to coordinate the efforts of the army, air force and marines in Central Colombia. Omega, the first of thirteen JTFs, continued to operate throughout the Uribe administration (and continued to exist until mid-2015, when it was disbanded in the face of defence budget cuts in the context of the peace process). The JTF's operations—which regained control of a large section of previously FARC-held territory in the Macarena region—created the conditions necessary for striking the FARC leadership. On 1 September 2007, Negro Acacio, head of FARC's 16th Front, which was heavily involved in drug trafficking, was killed by a precision air strike. In October, Martin Caballero, an important leader of the Caribbean Bloc, was killed. Over time, other similarly important mid-level commanders were killed or captured.

As well as disrupting its leadership, these operations also forced FARC to shift its geographical centre of gravity. Where it had been centred on the departments of Caqueta and Western Meta, it now shifted further southwest to Putumayo, Nariño, Cauca and Choco along Colombia's Pacific coast. Other FARC concentrations continued in the Sierra de la Macarena, Catatumbo and Arauca.

FARC was so disrupted that the leadership could not meet together in person—they had to meet virtually for their Ninth Conference. FARC leaders who participated in this conference admitted that their economic structures had been affected, and that the strategic plan had been disrupted and would take longer to achieve. Marulanda admitted that it would take four years of work for the organisation to recover its pre-Uribe levels of activity and organisation. Marulanda did not really have any new answers to confront the government, however; just the same formula of guerrilla warfare, mines and terrorism.[54] Given the difficult military situation, FARC focused on political efforts to mobilise all of the opposition to President Uribe and develop relations with foreign governments.[55]

On 1 March 2008, the Colombian armed forces and police bombed a camp just across the border with Ecuador, killing Raúl Reyes, a member of the

FARC Secretariat. In quick succession, a bodyguard killed FARC Secretariat member Iván Ríos, and Manuel Marulanda died of a heart attack at the age of seventy-seven while fleeing from an army attack in the jungle. While some began to predict the imminent collapse or fragmentation of FARC, this did not happen. Within days, all three members were replaced, with Alfonso Cano becoming the new head of the Secretariat. However, the degree of weakness of FARC due to the loss of leadership became apparent when intelligence units were able to infiltrate FARC communications and carry out one of the most spectacular rescue operations in history, Operation Jaque, liberating fifteen high-value hostages—including former Senator and presidential candidate Ingrid Betancourt, and three Americans who had been in FARC hands since 2003—without firing a shot and with no loss of life.[56]

Computers captured from Raúl Reyes's hideout in Ecuador revealed the previously unknown extent of FARC's international relations, exposing three levels of overseas relationships: foreign governments, foreign political parties and individuals, and foreign terrorist organisations. These relationships were formed to give FARC strategic depth to avoid defeat in Colombia. Particularly damaging were the relationships these documents revealed with President Hugo Chavez's Venezuelan government and President Rafael Correa's government in Ecuador.[57] The damage to Venezuela's reputation was reinforced in 2010 when Colombia presented significant evidence to the General Assembly of the Organisation of American States concerning FARC presence in Venezuela that was known, tolerated and even encouraged by the Venezuelan government. FARC, as explained earlier, had set up the COMINTER in 1993 at its Eighth Conference, and had ordered additional international efforts in its 2000 plenum, which set up the structure for subsequent international engagement. From 2004 onward, FARC had promoted the creation of the Bolivarian Continental Coordinator (Coordinadora Continental Bolivariana), composed of radical leftist organisations from Argentina, Brazil, Bolivia, Chile, Venezuela, El Salvador, Mexico, the Dominican Republic, Haiti, Puerto Rico, Cuba, Panama, Paraguay, Uruguay, Peru and Chile, as well as Spain, France, Italy and Denmark. Its major task was to develop solidarity and foment propaganda activity against the Colombian government and US policy.[58] The documents also revealed FARC's relationships with regional terrorist organisations including the Peruvian revolutionary movement, the MRTA; Venezuelan guerrilla groups; the Basque separatist group, ETA; and the Paraguayan communist guerrilla group, the EPP.

Phase six: 2008–present

Since 2008, the conflict in Colombia has been characterised by FARC attempts to adapt to, and overcome, government strength. The peace talks, currently ongoing in Havana, are part of this effort and are discussed in detail in Chapter 5.

In August 2008, Alfonso Cano produced a document that became known as 'Revolutionary Rebirth of the Masses' ('Plan Renacer Revolucionario de las Masas', often shortened to 'Plan Renacer'). While some called this a new strategy, in fact Cano's document confirmed FARC's continued will to implement its existing strategy, albeit using different methods. The main shift was a call for the guerrillas to return to basics. Where Marulanda in 2003 had advocated a return to guerrilla warfare as a temporary tactical measure, Cano insisted that the organisation return to guerrilla warfare as a strategic phase.[59] Most importantly, Plan Renacer called for political changes. Cano acknowledged that the organisation had 'misused our social resources and lost social-political space'.[60] FARC was to emphasise the growth of the Bolivarian Movement and the militias, and to work to create a legal political organisation called the 'People's Party' that would develop an alliance with the Bolivarian Movement.[61] Plan Renacer also confirmed the importance of the international relationships discovered in 2008, in particular with Venezuela.[62] By the election of President Juan Manuel Santos in 2010, Plan Renacer was in full force.

In September 2010, the Colombian Air Force and special operations forces carried out Operation Sodoma, killing FARC Secretariat member Jorge Briceño Suarez (alias 'Mono Jojoy'). Politically, this was a huge blow as Jojoy had been FARC's primary military commander since 1996. Jojoy's death certainly signalled the end of an era. However, it did not disrupt Plan Renacer, which continues at the time of writing.

While FARC was certainly severely weakened by 2010, it was not defeated and the war continued. Drug trafficking became increasingly central to the organisation's existence. FARC fronts acquired an increasingly dominant role in the drug business vis-à-vis other criminal and insurgent groups, expanding their participation to the transnational level and developing fronts that were exclusively dedicated to trafficking. They developed alliances with drug traffickers, and even with BACRIM. By 2010, as many as 45 per cent of FARC formations were known to be involved in one or more stages of the drug-trafficking business, from the most basic to the most advanced and sophisticated parts of the activity.[63] Their presence in all of the areas of drug growing, production and trafficking gave them access to every level of the illicit business.

They became major coordinators of the interconnected networks across the Americas, used to facilitate the shipment and international distribution of illicit drugs—ensuring that they held a dominant position in the trade. They sent representatives to Central America and Mexico, where they co-ordinated shipments with international traffickers.

In return, international traffickers frequently visited FARC-dominated areas in Colombia to verify the quality of the product and negotiate terms of sale, delivery schedules, and methods. The international trafficking of drugs to Central America and Mexico was coordinated through seven guerrilla units that established all the contacts, and coordinated the routes and the different means by which drugs were transported. The 30th, 48th, 57th and 58th Fronts largely favoured maritime routes, using semi-submersibles and fast boats, and hiding narcotics consignments in shipping containers. Meanwhile the 1st, 16th and 10th Fronts sent most of their consignments by air, using clandestine air-strips located in Colombia's eastern plains along the border with Venezuela.[64]

When he took over from Uribe, President Santos began to shift the empha-sis from security towards the economy—thinking, perhaps, that the war was on a downward trajectory. Management of the conflict was delegated to the defence ministry. However, complacency—both civilian and military—grew following the string of military victories and high-profile successes since 2003. FARC, employing the methodology Alfonso Cano had set out in Plan Renacer, took advantage of this to begin carrying out unexpected and lethal attacks that inflicted serious losses on the security forces. Development efforts in the problem zones—contested areas of the country, which had been over-seen by Uribe himself—began to decline with the reduction in accountability. In short, for the government things had begun to go backward.

President Santos dealt with this in two ways. First, he fired his first minister of defence, and appointed his trusted former chief of staff, Juan Carlos Pinzón, instead. Pinzón was the son of an army colonel, and was a highly educated economist who had served in the Uribe administration as a vice minister of defence. He was trusted by Santos, liked by the military and respected by the political opposition. Pinzón threw immense intellectual and organisational energy into the development of a new campaign plan to deal with the increased FARC violence, which became known as 'Espada de Honor' ('Sword of Honour'). Within sixty days of Pinzón's appointment, an intensive planning effort was completed and the military began to implement the plan in November of 2011, earlier than the official start date of January 2012. An early victory for Sword of Honour, and a good omen—even if it was the fruit of previous work—was the killing of FARC General Secretary Alfonso Cano on 4 November 2011 in Operation Odiseo.

The Sword of Honour campaign was built around mobile joint task forces—an expansion of the JTF structure that originated with JTF Omega in the Macarena region—to attack the areas of greatest FARC persistence, activity and violence. Ten key base areas were identified though a deep-dive analysis of FARC's 'rival system'. Three of these areas were already covered by existing task forces: JTF Omega on the borders of Meta and Caquetá departments; JTF Nudo del Paramillo in the Paramillo Massif, an incredibly rugged jungle-covered area at the northern end of the western Cordillera in the north of Antioquia; and JTF Southern Tolima, which had been doggedly pursuing Alfonso Cano through the Central Cordillera since 2008 and had just succeeded in killing him. The operational results achieved by these task forces had been positive and the military hoped to replicate this success through the creation of additional task forces. Although initially conceived of as joint formations, in reality these new JTFs rapidly devolved into service-based task forces subordinated to local area commands. Army task forces were subordinated to army divisions, and the navy and air force task forces were subordinated to their respective service territorial commands.

Due to resource constraints, it was decided to cover six of the seven remaining areas with new task forces. The army created four: JTF Vulcano for the Catatumbo area in North Santander, JTF Quiron covering Arauca, JTF Apolo to operate in Cauca, and JTF Pegaso for the department of Nariño. The air force created Task Force Ares, to suppress drug trafficking in Vichada; and the navy created Task Force Poseidon to cover drug trafficking out of southwestern Colombia along the Pacific coast from Nariño to Choco. JTF Southern Tolima was renamed Task Force Zeus under the new plan. In January 2012, all the task forces were short of personnel and equipment, reflecting the way that they had been cobbled together by stealing men and materiel from other commands. While still conducting combat operations, much of the first year of the Sword of Honour campaign was therefore, of necessity, spent building the JTF tables of organisation and equipment.

After a year, Pinzón formed a special planning team to review progress in the campaign: the result was an adjustment to the plan, now known as Sword of Honour II, based on experience and lessons learned. This new iteration of the plan included two additional Task Forces: JTF Titan to operate in the Choco, and JTF Algeciras to target FARC's elite Teofilo Forero Castro Mobile Column—a powerful grouping that fielded guerrilla forces in Huila and Caquetá departments and conducted urban terrorism activity throughout much of Colombia. There was also more emphasis on military-led stabilisation efforts to consolidate the terrain recovered to date. Sword of Honour and its successor were successful in suppressing much of FARC's violence. However, both operations were hampered by a chronic insufficiency of forces, transportation and logistics to carry out such an ambitious plan across the entire country. Furthermore, FARC adaptation—particularly the extensive use of mines and IEDs—slowed the military down considerably and inflicted grievous casualties. Although the military also inflicted serious casualties on all of the FARC fronts in the JTF areas of operation—in many cases virtually annihilating FARC fronts by inflicting almost 100 per cent casualties—the guerrillas also showed remarkable resilience, demonstrating a capacity to replace these losses almost completely. Thus, clearly, while the Sword of Honour campaign continues to go well militarily, there remain serious shortfalls in the kinds of social, political and economic advances that are essential to stem the numbers of young recruits willing to join the guerrillas.

At the same time that President Santos was encouraging Pinzón to increase operations against FARC and suppress the new rise in violence, he was also carrying out secret discussions with FARC through his peace commissioner, Sergio Jaramillo. Jaramillo, like Pinzón, was a close adviser to Santos who had worked for him as a vice minister of defence when Santos was defence minister under President Uribe. These secret discussions bore fruit when the government and FARC agreed to carry out formal peace talks in Havana on an agenda of a limited number of items over a one-year period. The agreed items for discussion were political participation, illicit drugs, victims, the end of the conflict, and implementation of any agreement. Norway and Cuba agreed to serve as guarantors of the process. The peace talks began on 19 November 2013 in Oslo, Norway and then moved to Havana, Cuba.

At the time of writing, peace talks are approaching the end of their second year. While progress has been made, much uncertainty remains. FARC leaders do not want to give up their weapons—preferring merely to stop using them—and, to date, are refusing to consider any sort of jail time or other punishment for any crimes committed in the name of insurgency. While there does seem to be a general intent by FARC to sign a peace agreement, what that agreement will look like, and whether it will be viable, remain open questions.

Conclusions

As this necessarily brief historical survey demonstrates, the combination of longstanding patterns of weak central government, a culture of centralism (of always looking to Bogotá), abundant natural resources and broken terrain has shaped the history of FARC, exacerbated the problems of Colombia's periphery, and driven the conflict. (These factors are discussed in more detail in our analysis of guerrilla and counter-guerrilla warfare in Colombia in Chapter 3). What changed from 1994 onward was that, as FARC launched its 'war of movement' against Colombia's main urban centres, the problems of the periphery finally began threatening to overpower and overwhelm the centre. This backfired for FARC, however, as Colombians unified and fought back under President Uribe.

The problem for the government is that these very security successes may prove their own undoing. As the threat to the centre from the periphery has diminished, so has the centre's interest in fully solving the problems of the periphery. Colombia is returning to politics-as-usual and this, in large measure, may explain why the government has been anxious to cut a deal with FARC. The quicker FARC goes away, the quicker Colombians can focus again on the centre, even though this may be a very dangerous proposition. Even if FARC does go away, unless the problems of the periphery are solved, some other enemy of the centre—the child or grandchild of FARC—will fill the vacuum, and conflict of some sort will continue in Colombia.

2

BUILDING THE TOOLS FOR MILITARY SUCCESS

Dickie Davis with *Anthony Arnott* [1]

War has always driven military change and adaptation, and the history of Colombia's armed forces, as they developed through the many phases of conflict described in the last chapter, is no different. In this chapter, we explore Colombian military adaptation through the lens of capability development—examining how the Colombian military identified and acquired the people, equipment, doctrine, training and infrastructure vital for success.

Colombia's war with Peru from September 1932 to May 1933 demonstrates perfectly the way that military necessity—often in the guise of unexpected setbacks or defeats—drives capability development. In 1932, President Luis Miguel Sánchez of Peru invaded Colombia with two army regiments sent to Leticia and Tarapacá, both in a remote area of the Colombian Amazon. He believed that Colombia had little chance of defending itself due to the inaccessibility of the region and Colombia's lack of air force and navy—the latter having been disbanded in 1909. The Colombian response was swift: within ninety days, a fleet of old European ships had been assembled and had reached the mouth of the Amazon. Meanwhile the small military aviation fleet of five aircraft was rapidly expanded with the purchase of seventy-four additional planes and the loan of pilots and aircraft from the Colombian–German Air Transport Society. Main combat operations began on 14 February 1933 and fighting was brief, with the last military actions in the conflict finishing on 8 May 1933. The conflict was finally resolved by the League of Nations, which reinstated the existing border. Perhaps more importantly, the Colombian Navy and Air Force were changed forever as a result.

For militaries to develop new equipment and capability during counterinsurgency (COIN) operations is not uncommon. The United States and its allies, for example, saw significant advances in unmanned aerial vehicles (UAVs), technology to counter improvised explosive devices (IEDs), and armoured vehicle design during their campaigns in Iraq and Afghanistan of the 2000s. The UK likewise developed countless new capabilities during the four decades of The Troubles in Northern Ireland—from sophisticated electronic counter measures to cutting-edge surveillance equipment on the ground and in the air.

Colombia's campaign against FARC, the ELN, BACRIM and other groups has been no exception in this regard. At the start of the conflict, Colombia's armed forces were designed to protect the country from external threats. This had big implications: force design affects everything from troop numbers to the location, number and size of bases; weapons and communications systems; the security force's core organisational structures; and command and control arrangements. From the early 1960s until the late 1990s, a series of incremental changes were made to all three armed services and the National Police. As the conflict was primarily regarded as a police and army problem, most adjustments were made to these organisations. However, the case for reform became less pressing after 1982, as the government's overarching approach became one of attempting to resolve the conflict through a peace process.

The serious tactical setbacks of 1997 and 1998 (described in Chapter 1) convinced the government and the leadership of the armed forces that more radical change was needed to modernise and strengthen both the military and the police. This change was driven by a new view of how operations should be conducted. The role of ground-holding, territorial forces was to remain, but the new vision put increased emphasis on intelligence collection and processing, improved communications, the development of mobile forces that could focus on and strike the enemy, and on better civil-military integration. The tactical setbacks also forced a change of mind-set: this was no longer just a police and army problem; all elements of the Colombian armed forces became focused on the defeat of the guerrilla groups.

The changes required were huge. The security forces needed to undergo a rapid expansion to generate the numbers of troops required to deliver effective security. The whole equipment programme needed to be refocused to deliver the mobility, weapons, communications and support systems needed for successful COIN—for example, the introduction of better-armoured vehicles to protect against mines, and a vastly improved riverine capability for naval operations on Colombia's 11,000 kilometres of rivers. Bases needed to be reconfigured across the country to ensure, for example, that the air force had proper coverage of the whole country and that increased force numbers could be accommodated. Training needed to be revamped to deal with changing guerrilla tactics and techniques and with emerging human-rights issues. The structures and roles of task forces needed review. Logistic capability needed to be increased to support the enlarged force for operations in the jungle. These changes needed to take place across the army, navy, air force and police, and in joint units.

In late August 2002, President Álvaro Uribe was informed by his finance minister that he did not have enough money to pay the army beyond the end of October 2002.[2] Three weeks into his term, this came as something of a shock. The 1999 Colombian recession had been the worst in seven decades and the conditions of the $2.7 billion International Monetary Fund bailout had committed the government to structural reforms and budget discipline. For the new president, two months of hard work followed to find a path through the crisis. Part of the answer was the creation of a new 'democratic security tax'—initially set as a 1.2 per cent levy on liquid assets,[3] paid by the wealthiest taxpayers (both individuals and corporations) to be used for the expansion of the army and police.[4] Though nervous about the effects of such a tax, Uribe agreed to the idea because it avoided cost-cutting measures in other areas to fund the expansion, and it would show that the wealthy—long criticised for their indifference—were willing to sacrifice to make their country safer.[5]

Driven to a large extent by a requirement to keep the cost of the armed forces (which, in Colombia, include the National Police) affordable, the Colombian Ministry of Defence has focused not on the most advanced but rather on the most cost-effective equipment. Thus the security forces have been built with a ruthless pragmatism, focusing on the problem to be solved and a realistic approach based on the money available. As a result of the ensuing changes, Colombia now possesses a security force that is highly competent in jungle COIN, with a real depth of operational experience and a unique combination of capabilities. The effort that has gone into creating this 400,000-strong force is one of the untold stories of the campaign.

It is impossible to cover all the changes in complete detail in just one chapter. However, to illustrate the scale and complexity of the changes, we have chosen to describe how the Colombian armed forces developed their capability to capture or kill high-value targets—a capability which, as described in the previous chapter, inflicted massive damage on FARC's senior leadership in 2007–08, forcing the rebels to abandon their 'war of movement' and drop back to the guerrilla-warfare stage in their Protracted People's War strategy. This has thus been a decisive capability—one that has required a truly joint approach, and ultimately has made headlines around the world.

High-value targets: combining intelligence, air power and special operations

On the evening of 22 September 2010, the jungle canopy of Meta was breached by an assault that would last several days and kill one of FARC's most notorious leaders. Mono Jojoy—once described by President Andrés Pastrana as 'the real narco-terrorist'—was known for his brutal methods, being the leading military commander of FARC, and a contender to become overall leader of the organisation. Approximately sixty aircraft were involved in the operation, including thirty fixed-wing planes and twenty-seven helicopters, dropping an estimated 3 tons of explosives. The killing of Mono Jojoy in his jungle hideout near La Macarena was a dramatic blow to the leadership of FARC, but was even more so a demonstration of a massively enhanced Colombian government capability. The operation combined air power, technical and human intelligence, and special operations forces into a complex joint operation involving the army, air force and police.

In the weeks before the operation, a GPS tracking device was placed into the heel of Mono Jojoy's boot. One of the effects of his diabetes meant he needed special footwear—an almost literal Achilles heel. In addition to the GPS tracker, a human source with eyes on the target is reported to have confirmed Jojoy's position. Armed with this intelligence and coordinated by the relatively new Joint Special Operations Command (CCOES), a formation of A-29 Super Tucanos, supported by the Israeli version of the Mirage 5, the Kfir, bombed Jojoy's camp. GPS-guided munitions would have been released from several miles away at an altitude of 20,000 feet, delivering the first stealthy blow in a major operation targeting the heart of the guerrilla organisation.

The follow up was a helicopter assault led by the air force, but also incorporating police and army helicopters. Since police helicopter crews did not have experience of fast-roping—a method of getting air assault troops out of a hovering helicopter using fixed large-diameter ropes—air force and army crewmen were lent to the National Police for the operation to provide a short-term solution. This was a testament to the level of integration to which the services had become accustomed.

Apart from killing one of FARC's most senior leaders, the operation also killed between twenty and thirty rebels and led to the capture of over 100 computer hard drives and memory sticks, yielding a huge quantity of data. The biblically named Operation Sodoma also provides an example of how effective the Colombian armed forces had become by 2010. The operation fused sophisticated technological and human intelligence. It involved exercising command and control across three different services. It also included a complex air operation, using the relatively newly attained skills of precision-guided attack and the well-honed use of night-vision goggles, which had started in the late 1990s.

Yet fifteen years earlier, the picture had looked very different. The Colombian Air Force at that time was largely equipped to defend the country from outside aggressors, not from those within. The Aviación del Ejército— the army's aviation branch, distinct from the Colombian Air Force—only came into being in 1995, and police aviation was also limited. Intelligence was handled very differently too. General Jorge Hernando Nieto Rojas, a police officer who is Colombia's head of public security, believes that the main change in thirty-five years has been in intelligence sharing. The services used to keep intelligence to themselves, but today there is a wider understanding of the unparalleled effectiveness of all-source intelligence fusion and close sharing among branches of the security forces. 'Now', he reflected, in 2015, 'he who has the information ... and knows how to use it has the power.'[6]

As Jim Rochlin points out, early reverses in the campaign showed that 'Colombian forces needed to be quicker, more mobile, equipped with better intelligence, more capable of fighting in difficult terrain such as high mountains and also rivers.'[7] In Operation Sodoma, as in other high-value-target operations, the Colombian armed forces showed they were developing these capabilities, and were bringing together impressive intelligence, airpower and special operations capabilities. The improvements in these three capability areas in particular were so significant that they are worthy of more detailed examination.

Intelligence

The Colombians had long realised the importance of intelligence in their battle with the various guerrilla groups plaguing their country. There is a considerable history of reforms to various parts of the intelligence system—not least the complete disbandment and replacement in 2011 of the scandal-plagued national intelligence agency, the Departamento Administrativo de Seguridad (Administrative Department of Security, DAS), by President Juan Manuel Santos. Speaking as defence minister in 2009, and at odds with President Uribe's view, he told journalists that DAS 'is an illness in a terminal state that should be given a Christian burial'.[8] When he became president he enacted his promise to disband it.

Even as early as the 1950s, returning veterans who had served in Colombia's contingent in the Korean War could see the need for intelligence reform and started to argue for changes. US efforts in support of reform began in 1961 with the deployment of a two-man military-intelligence training team to Colombia. Perhaps the most significant early reform involved recognition of military intelligence as a discipline in its own right, with the formation of the Intelligence Corps in May 1991. Before this, intelligence was centralised and directed from Bogotá, but each service acted independently with little cross coordination. Furthermore, intelligence was collected almost entirely from human sources, making the accurate location of time-sensitive targets, such as the leaders of guerrilla groups, challenging. One commander described the situation: 'Before 1991 we were blind, a second lieutenant was looking for the FARC in the jungle with his platoon on his own; from 2002 we started to get decent intelligence'.[9]

After May 1991 specialised military intelligence units were created and allocated to the regions. However, it was the location and killing by the police of narcotics kingpin Pablo Escobar in December 1993 that demonstrated what was possible. Towards the end of his life, Escobar was increasingly isolated and forced to communicate with his family and followers by mobile phone and radio. He was finally located using signals intercepts and direction finding, confirmed by physical observation. This success showed what could be achieved if these capabilities could be replicated against the leaders of the guerrilla groups. Over the mid-to-late 1990s, the army created an organisation to focus its intelligence assets on high-value targets, and some progress was made on locating leaders.

Although human assets were to remain the most important source of information, the armed services and the police needed to acquire more sophisticated means of locating and tracking their targets and gathering intelligence. It was here that US support was vital. Plan Colombia, conceived between 1998 and 1999, included $31 million for counter-narcotics intelligence, including land radar systems and the associated command and control systems to better track incursions of air space. Additionally, it included $17.4 million for aerial counter-narcotics intelligence. The media has also made numerous claims of more funding being made available through a 'black budget'.[10] Given the overlap between the guerrillas and drug traffickers, this support was useful for many things. For example, satellite images with resolutions of between 1 and 9 metres used for crop recognition could not necessarily differentiate between a guerrilla and a local civilian—but they could allow for the identification of guerrillas operating in large groups, and detect their bases when not obscured by clouds or jungle canopy. Advances in equipment did not just come from the United States: Colombia has devoted considerable national resources to improving its intelligence-collection capability including, for example, the introduction in 2013 of the Israeli Elbit Hermes 900 and 450 UAVs—small, agile, remotely-piloted surveillance drones. But US support was vital in deploying and operating sophisticated communications intercept, positioning, and other technical surveillance capabilities, as well as for access to wider US intelligence sources.

One of the key challenges for intelligence is rapid, secure dissemination across the force. To solve this issue, the United States allowed Colombia access to its Combined Enterprise Regional Information Exchange System (CENTRIX), a coalition data-sharing network. This meant that intelligence could be disseminated rapidly, interrogated by those in the field who needed it, and a common operating picture established.

When elected in 2002, President Uribe, acting on his belief in citizen security (a key element of the DSP), instituted an aggressive reward scheme for those who provided useful intelligence:

> We started another key initiative named Lunes de Recompensa, or 'Reward Mondays'. The program was self-explanatory—every Monday, in villages and cities throughout Colombia, we paid small cash rewards to citizens in return for critical information on the movements of terrorists, petty crimes and other useful information ... By the end of our presidency, we ended up with more than four million people nationwide ... providing information in some fashion to the state.[11]

This produced lots of information, but also made guerrilla groups wary of infiltration. Human intelligence remains one of the key sources for high-value target (HVT) operations, producing the traditional dilemma of needing to protect sources, yet giving those using the information confidence in its accuracy. Such was the fear of compromise that strict controls were put in place to protect the intelligence being gathered: for example, those tasked with attacking the target only found out the identity of the HVT once they were deployed.

President Uribe also gave instructions to the military to go after FARC leaders and capture or kill them. This order energised the progress resulting in the creation of a Joint Operational Intelligence Committee (JOEC), a serious attempt to break down inter-service barriers and get all parties to share their material.

However, by 2006 there was considerable frustration amongst the Colombian military leadership at the lack of results for their efforts. It was at this point that the government of Colombia received valuable methodological support from former Israeli officers, which was, in many ways, the key to the considerable successes eventually delivered.[12] The Israelis taught Colombian officers target-centric intelligence fusion—a methodology in which each HVT becomes the focus of a joint team, with the service with the most developed information leading the team. The team leader, normally a colonel, would be given considerable authority and a high priority for tasking (that is, obtaining exclusive support from) scarce intelligence collection assets. A key factor was that a member of the operational team that would eventually plan the attack on the target was embedded in the joint intelligence team from the beginning. The teams worked hard to build comprehensive 'patterns of life' for their targets, such that they could begin to predict their target's future actions and likely responses to events. The Israeli support was short-lived, but its effect was huge for it finally brought everything together. The results began to be seen in March 2008 with the deaths of Raúl Reyes and Iván Ríos.

In the press releases following these two major successes, the Uribe administration claimed that aid provided under Plan Colombia had dramatically improved the government's intelligence-gathering capabilities. Indeed, Colombian military officials acknowledged that an intercepted satellite telephone call from Hugo Chávez to Raúl Reyes had helped locate the latter on the border with Ecuador. The problem, however, was that these and other press releases revealed too much information. Guerrilla leaders learnt that a key part of the targeting was the use of human sources within their inner circles. As a result, they changed their personal-security arrangements and reduced the numbers of their close contacts—though this in turn made it harder for them to control their organisations.

The effect of the loss of key members of the guerrilla leadership has been hotly debated. Some have argued that this had the potential to make the organisations more radical and made it harder to have a sustained dialogue. But whatever the effect on these organisations, it gave the government visible, high profile successes, creating momentum and puncturing the myth of invincibility, particularly of FARC. From this viewpoint, these missions were a huge success; they signalled the turning of the tide.

The challenge now facing Colombia's intelligence organisations is one of knowing how to reform in response to the changing situation on the ground. These organisations were built to deal with a specific problem, and the nature of that problem is now changing. As one army officer put it, 'the intelligence guys have had a long time to study the FARC and the ELN; we know a huge amount of detail'. If the guerrillas lay down their arms and enter the political arena, a considerable shift in intelligence thinking and approach will be required. That process is already under way in the Directorate of Police Intelligence, which is beginning the move away from COIN to focus on citizen security. Reflecting on Colombia's successes and Mexico's challenges, a senior Colombian police intelligence officer remarked, 'Our strength is our structure and the fact that we have been made to work together for the national good. In Mexico the police force is not integrated, they don't talk to each other; we used to be like that'.

Air power

At the beginning of the 1990s, the Colombian Air Force was a relatively small organisation containing only 8 per cent of Colombia's military personnel. Its roles were largely focused externally—on being able to protect its skies against a regional aggressor. Its mainstay in this task was the Israeli-made Kfir, armed with Python air-to-air missiles. The air force had a limited helicopter capability in the form of a mixed fleet of the ubiquitous Bell UH-1 Huey—of which the country had fewer than twenty—and a handful of Hughes 500s, used for a combination of training and liaison tasks. The air force's design therefore meant that it was not able to deal with the growing narcotics issues beyond a limited ability to intercept some of the roughly 500–1,000 narcotics-smuggling flights per year. Yet by the late 1990s, the air force had not only grown, but tailored its capabilities to face the increasingly critical threats of guerrilla warfare and narco-trafficking.

As discussed in Chapter 1, in November 1998 over 1,000 FARC guerrillas captured Mitú, killing many police officers and taking the remainder of the small force hostage. With reinforcements more than 100 kilometres away, it was immediately apparent that any attempt to recapture the town was going to have to come from the sky. With no other airfield within 400 kilometres of Mitú, an airborne task force was assembled at a Brazilian air base just over the border. A mixed force of UH-60 Blackhawks, two AC-47 Spooky gunships, and two C-130 transport aircraft touched down at the base—met by a concerned and surprised Brazilian base commander, whose seniors had not relayed the inter-governmental agreement to him. Operating from this temporary airhead, the air force commenced operations around Mitú. The AC-47s provided an important picture of what was unfolding in the besieged town below, as well as fire support from their 0.50 calibre machine guns. The AC-47s—upgraded C-47 Skytrains—were popular with ground forces for their loiter time of six hours. This ad hoc task force went on to insert over 300 troops over the coming days, and to return law, order and governance to this remote Colombian town. The lessons from Mitú were as stark as they were simple: air power was going to play a key role in defeating the guerrillas. As one senior air force officer who flew during the operation put it, 'We realised that we needed to start talking to the Army'.[13]

It was not just the air force that was going to feature heavily in this air campaign. Colombia now holds of one of the largest fleets of helicopters (320) in the world, along with fixed-wing aircraft (265), refined for precision operations and surveillance tasks across the army, air force and police. The army aviation capability was formally established in August 1995 to provide logistic support to the army. The Servicio Aéreo de Policia (Police Aviation Service) also significantly bolstered its fleet, now operating approximately forty planes and sixty-five helicopters.

The rapid evolution in Colombia's air power has been one of considered pragmatism—the advances in its air capability have by no means solely been the consequence of increased funding. On the contrary, there are many stories of innovation and cost effectiveness, such as the Super Tucano aircraft, Arpía helicopter, and Spike missile.

The story behind the use of the turboprop A-29 Super Tucano light attack aircraft is illuminating. As one senior Colombian Air Force officer reflected, 'we needed a jet attack aircraft, but we were given the A-29'. This was a financial decision. But the result has been far from low-tech or low-performance. Purchased in 2006, the Super Tucano was well adapted to the campaign. Competent crews could drop their Mk 82 unguided bombs onto a 10 square metre area. Whilst the engine noise of the relatively low and slow-flying plane could sacrifice the element of surprise, it could also act as a psychological weapon. According to interviews with former FARC militants, fear of bomb-ing was one of the main reasons for their desertion;[14] furthermore, that they could often hear the planes flying around at night was enough to keep them lying awake in their trenches for hours on end, leading to lasting fatigue and weariness. Later versions of the A-29 were equipped with laser rangefinders, which proved essential in the early use of Paveway II laser-guided bombs deliv-ered by A-37 Dragonflies. At less than $2.5 million per aircraft, the Super Tucano has proven to be a highly cost-effective COIN platform for the Colombian Air Force.

Although Colombia procured its first UH-60 Blackhawk in 1989, it was not until 1995 that the Arpía variant was born. Named after the South American harpy eagle (known for preying on tree-dwelling monkeys), it was an apt name for Colombia's guerrilla-fighting attack helicopter. Since 1995 the aircraft has been through four stages of development; the Arpía 4 is a formidable attack helicopter by any measure. It can be equipped with any combination of miniguns, rockets, or the Spike anti-tank missile on its exter-nal weapons pylons. It also has two crew-manned 20-mm miniguns mounted in order to give added firepower, especially when the aircraft is climbing away from the target.

Whilst the UH-60 and its utility, attack, and casualty-evacuation variants may have been the backbone of Colombia's helicopter force, it has not been the only platform. Colombia operates sixty-two Bell UH-1 and Bell 212 heli-copters, and the army operates sixty-four similar variants. Much of Colombia sits high in the mountains, the altitude proving a challenge for many types of helicopters. In 1997, Colombia purchased ten Russian Mi-17 helicopters, robust aircraft well known for their performance at altitude, proven during the Soviet Union's war in Afghanistan.

The Spike missile is worthy of particular note. Developed in Israel, the weapon can be launched from up to 24 kilometres from its target, is guided by GPS, and in the latter stages of flight an Arpía crew have the ability to fine-tune the target—for example, to ensure it enters the target building through a particular window.

Yet the Spike missile was not the Colombians' first precision weapon; indeed, it is only a relatively recent addition to the arsenal. In 2006, the then Defence Minister Juan Manuel Santos asked US Secretary of Defence Donald Rumsfeld for some precision-guided munitions—specifically, the GBU-12 Paveway II. The Paveway II is a Mk 82 500-pound general-purpose bomb with an added guidance package that turns it into a laser-guided weapon. This represented a technological quantum leap—described by Juan Carlos Pinzón as a 'turning point'—from a circular error probable (CEP, a measure of accuracy) of about 120 metres, to one of less than 4. So successful was the trial, with the first bomb landing within 30 centimetres of the target, that it was immediately taken into service and fitted to the A-37 Dragonflies.[15]

In September 2007, FARC's chief drug trafficker, Tomás Medina Caracas (alias 'Negro Acacio'), was the first Colombian target of these precision munitions. It was the Paveway II that was to bring about the demise of not just Mono Jojoy, but also Raúl Reyes, whose Ecuadorian compound was destroyed by a bomb dropped from inside Colombian airspace. Other members of the FARC leadership have experienced the same fate. It was not just a smart bomb, but smart procurement. The cost of each Paveway II is $30,000—in contrast, Hellfire missiles, used widely in Iraq and Afghanistan, cost $75,000 each, and the Paveway III costs $120,000.

It is not only equipment that has changed Colombia's use of air power, however; so too have developments in doctrine and training. On 13 December 1998, a Bell UH-1 operated by the air force is alleged to have dropped a cluster bomb in the centre of the town of Santa Domingo. The account of the events remains contested, but it is suggested that seventeen civilians were killed in this tragic episode. Whether it was an air force cluster bomb, or another bomb detonated by guerrillas, may be a question that is never resolved. Nevertheless, the aftershocks altered the way the air force operated. Prior to the Santo Domingo incident, crews would not even file a report if they had opened fire. But Santo Domingo 'changed many things', reported one air force pilot: 'it impacted our planning, including incorporating lawyers into the planning process. It made us realise that we needed to be like a surgeon'. In fact, the incorporation of legal opinion during the planning and targeting process is a change that has been reflected across all of Colombia's armed forces—not just those operations executed from the skies.

The importance of air power in COIN and similar operations is well documented. The Soviet Union's eventual widespread use of helicopters in Afghanistan marked a turning point in its favour—and the Mujahidin's use of shoulder-launched anti-aircraft missiles was one of the first markers of the reversal of that success. The expansion of Colombia's air power, along with its application of low-cost precision-guided munitions, has been critical in the success of the campaign. The considered equipment-procurement processes have enabled Colombia to expand its capabilities, and numbers, without breaking the bank.

Special operations

Special operations capabilities—including long-range patrolling, close target reconnaissance, hostage rescue and recovery, raiding, unconventional warfare (the raising and employment of local irregular forces in a hostile environment), direct action against HVTs, and certain forms of advanced special intelligence—have long been recognised as a fundamental element of counter-guerrilla warfare (the military subset of COIN, discussed in detail in Chapter 3). As a result, governments engaged in COIN since the mid-twentieth century have typically invested significant effort into developing these capabilities, and Colombia has been no exception.

The Colombian special operations forces were established over forty years ago with the re-purposing of an infantry unit that became known as the Special Forces Rifle Battalion. Over time, this battalion moved back into long-term COIN missions; as a result, the army created, in the late 1970s, the Special Forces Group—an organisation of between sixty and eighty officers and non-commissioned officers (NCOs) who, in effect, formed a permanent special-operations cadre. Members of this group were organised and trained as specialists and grouped together in inter-disciplinary teams. By 1980 the unit had gained a reputation as the most elite force in the army due to considerable operational successes against the urban-oriented M-19 terrorist group. In the words of one former member, 'we did not have the equipment, doctrine or collective training, but the individual training and the quality of the soldiers was outstanding'.[16]

But in 1981, in an operation against FARC in the department of Caquetá, the unit was ambushed in dense jungle and took heavy casualties. This sparked a serious review of tactics, training and equipment—'why were the soldiers using MP5 machine guns, a weapon best suited for the urban environment, in the jungle?' As a result a decision was taken to retrain and expand the force, which until this time had been composed solely of officers and NCOs, with the addition of 100 enlisted soldiers drawn from long-service volunteer ('professional') troops rather than short-term conscripts. During the mid-to-late 1980s, the expanded unit was deployed around the country as a quick-reaction force, responding to crises and fire-fighting. Yet in the words of one senior Special Forces officer, 'the employment of the force was done by people who did not really understand how to use the unit effectively'. The force became known as the 1st Special Forces Group (Agrupación de Fuerzas Especiales) and was employed mainly in rural areas.

In 1985, the Palace of Justice—the seat of Colombia's supreme court—was seized by M-19 terrorists. During the subsequent siege and counter-attack, the rebels were killed, along with many security force personnel and eleven out of the twenty-five Supreme Court justices; in all, more than 100 people lost their lives. Following the siege, President Betancur appeared on national television and accepted responsibility for what he described as a 'terrible nightmare'. As a direct result of this event the government, with the support of the US, decided to create a dedicated urban counter-terrorist capability. This unit— the Agrupación de Fuerzas Especiales Antiterroristas Urbanas (Urban Counter-Terrorism Special Forces Group, AFEUR)—has expanded over time and is now part of Joint Special Operations Command (CCOES).

In the early 1990s, the 1st Special Forces Group grew from a company-sized unit into a battalion and a second special forces battalion was created. In 1993, the two were combined under an army special operations headquarters, solving some command and control issues. Known as the Comando Operaciones Especiales de Contraguerrillas (Counter-Guerrilla Special Operations Command, COECG), this headquarters was first commanded by General Jorge Mora. Initially these battalions were used as quick-reaction and strike forces. Such was their success that, in 1995–96, two more battalions were created, initially by re-purposing existing counter-guerrilla battalions (specialist light infantry belonging to the army's general-purpose forces, organised in small, agile battalions of roughly 350 personnel). The United States supported this rapid growth with both equipment and training. Their special treatment and better equipment served to enhance the status of the special forces, attracting high-quality candidates into their ranks. These four battalions were organised into the 1st Special Forces Brigade (Brigada de Fuerzas Especiales), under the operational control of the army.

In 2003, when General Carlos Ospina Ovalle became commander of the army at the height of President Uribe's counter-offensive against FARC encroachment on Colombia's major cities, he soon saw that the 1st Special Forces Brigade was being employed as a quick-reaction force. He realised that a force was also needed to attack the enemy's rear, capable of undertaking the long-range, deep penetration missions needed to take the battle to FARC leadership. As a result, in 2003 he ordered the raising of Colombia's first com-mando battalion (Batallón de Commandos, BACOA). This move caused a degree of internal friction and resistance as existing battalions within the renamed 1st Special Forces Brigade wanted this new role, but the change was forced through. Commando selection was open to all in the army and was tough. BACOA was organised into two light-infantry companies and one reconnaissance company made up of six-man reconnaissance patrols, a configuration suited to jungle raiding and long-range missions.

The Colombian army had run a 'Lancero' course since the late 1950s—based on the United States Army Ranger School—to promote leadership and tactics for irregular and small-unit warfare. The term 'Lancero' (Lancer) commemorates an elite unit who fought for Simon Bolívar during Colombia's war of liberation in the nineteenth century; in its modern application, a Lancero is a soldier belonging to the general-purpose forces of the army (not a special operations soldier) who receives additional training in light infantry tactics, weapons and leadership, in a similar manner to the US Army Ranger School or the UK All Arms Commando Course. During La Violencia in the 1950s, the army created Lancero companies for raiding missions against guerrillas, but when the conflict ended these were disbanded. In 2004, they were recreated as the Agrupación de Lanceros (Lancero Group, AGLAN) for the purpose of deep jungle raiding.

BACOA and AGLAN were combined into a newly formed Comando de Operaciones Especiales-Ejercito (Army Special Operations Command, COESE) and, once they started to get good intelligence, began to carry out a series of raids against the FARC and ELN leaderships, scoring some early successes. In time COESE became the army contribution to the CCOES.

CCOES is a joint, 2,050-strong force responsible directly to the head of the armed forces. The organisation consists of five special-forces battalions (four army and one marine), the urban anti-terrorist unit (AFEUR), and a small air-force unit that focuses on battle-damage assessment following bombing raids. CCOES does not have its own air wing, but its missions are given a very high priority by the air force and army. Army aviation operates a special operations aviation battalion (Batallón de Operaciones Especiales de Aviación, BAOEA) in direct support. CCOES focuses primarily on specialist tasks associated with the HVT campaign undertaken since 2007, while the army has retained control of the 1st Special Forces Brigade, which remains focused on long-range patrolling, raiding and reconnaissance in a jungle environment. The navy and air force both retain smaller special operations capabilities. The capabilities and performance of Colombia's special forces are today regarded as world class—indicated in part by their impressive operational record and series of wins in the annual 'Fuerzas Commando' competition that brings special operations troops together from across Latin America.

Perhaps a less positive indication of the widely recognised quality of Colombian special forces is the fact that former members have been in extremely high demand as security contractors for private military companies supporting the wars in Iraq, Afghanistan and elsewhere since 2001—a 'brain drain' of which their leaders have had to be especially watchful.

Bringing it all together

Each of these three transformations of intelligence, air power and special operations is impressive in its own right, especially given the fact that they took place during a period of intense conflict, often against considerable resistance from the powerful individual services and the National Police. But much more important than each of these individual transformations has been the synergy—the integrated tactical, operational, and strategic effect—achieved by implementing all three capability transformations simultaneously, such that the improvement in effectiveness has been much more than the sum of its parts.

This overall transformation is about much more than just buying the right equipment or recruiting the right numbers of troops and police. It has been about the holistic development of capability across all lines of development[17] and integration between the three services and the police. Of the two, the integration across service boundaries has been the hardest to achieve, and has only been done through a ruthless focus on taking the fight to the enemy. This remains an ongoing and difficult challenge: as one experienced insider remarked, 'You need a strong fist to achieve integration. Half the integration story you hear is true; half is bullshit'.[18]

By June 2014, Mono Jojoy's camp outside La Macarena—the scene of his demise in 2010—had become a small special forces camp, displaying pictures of his time in the camp and of the hideous conditions in which his captives were held. It is a sad place. Hidden from the air by thick jungle and with plenty of tunnels and bunkers for protection from an air attack, it is clear that he was taken by surprise—a surprise that had been over eight years in the making.

3

GUERRILLA AND COUNTER-GUERRILLA
WARFARE IN COLOMBIA

David Kilcullen

This chapter examines guerrilla and counter-guerrilla warfare in Colombia. It builds on the historical account and discussion of capability development offered in the last two chapters, providing a military-technical analysis of FARC's operational system—the guerrillas' organisation, operations and tactics—and of the Colombian government's counter-guerrilla approach.[1]

It is argued that objective conditions in Colombia create an underlying 'territorial logic of stalemate'. This dynamic makes it virtually impossible for a rural guerrilla movement to overthrow the Colombian state, but also renders it almost equally difficult for governments to destroy guerrilla movements, as long as the guerrillas confine their operations to the country's vast, remote and sparsely populated periphery.

In this reading of the conflict, the fundamental strategic error of FARC leaders after 1993 was to transition to a 'war of movement', take the conflict to the cities, and attempt to defeat the government on its own turf in the industrialised, urbanised, heavily populated core of the country. This prompted a massive government counter-offensive that forced the guerrillas back into the periphery and onto the defensive—re-establishing the traditional equilibrium and leading to today's peace talks. But the government now confronts the same dynamic of stalemate in reverse: without long-term, sustained efforts to change the conditions in Colombia's periphery that have traditionally given rise to rural guerrilla warfare, the conflict will return.

The term 'counter-guerrilla warfare', used here, denotes military and police combat operations against a guerrilla enemy. By contrast, the broader concept of counterinsurgency (COIN) is 'a comprehensive civilian and military effort designed to simultaneously defeat and contain insurgency and address its root causes.'[2] COIN in this wider sense is a whole-of-government endeavour to defeat all armed and non-armed aspects of an insurgency. It includes political, economic, social, diplomatic, and informational action programmes, as well as security operations. Counter-guerrilla warfare, on the other hand, is 'geared to the active military element of the insurgent movement only'[3] and is conducted primarily by the military and certain types of police or intelligence units (collectively known as 'security forces'). Counter-guerrilla warfare is thus a subset of COIN, albeit a critical one since its goal is to create security—the prerequisite for every other counterinsurgent activity.[4]

As in all warfare, material conditions and subjective perceptions (which together form the geographical, demographic, economic, psychological and political context for any given instance of conflict) are hugely influential in determining the organisational and operational methodology each adversary adopts in a guerrilla war. Thus, before examining FARC's operational system and evaluating current issues within the rival counter-guerrilla system, it is worth briefly reviewing Colombia's overall conflict setting and the resulting territorial logic of stalemate.

Colombia: the domestic setting

In mid-July 1964, as survivors of the Marquetalía Republic stumbled out of the jungle-covered mountains of the Cordillera Central towards Rio Chiquito and took their first steps towards founding FARC, Lieutenant Colonel David Galula of the French colonial infantry was putting the finishing touches to his classic *Counterinsurgency Warfare: Theory and Practice*.[5] In his second chapter, Galula—whose experience extended to China, Greece, Algeria and Indochina, though not to Latin America—laid out the prerequisites for successful insurgency as he saw them. Among other factors, he emphasised geography, demographics and economics:

1. *Location.* A country isolated by natural barriers (sea, desert, forbidding mountain ranges) or situated among countries that oppose the insurgency is favourable to the counterinsurgent.
2. *Size.* The larger the country, the more difficult for a government to control it ...
3. *Configuration.* A country easy to compartmentalise hinders the insurgent ...
4. *International Borders.* The length of the borders, particularly if the neighbouring countries are sympathetic to the insurgents ... favours the insurgent.
5. *Terrain.* It helps the insurgent insofar as it is rugged and difficult, either because of mountains and swamps or because of the vegetation ...
6. *Climate.* Contrary to the general belief, harsh climates favour the counterinsurgent forces, which have, as a rule, better logistical and operational facilities.
7. *Population* ... The more inhabitants, the more difficult to control them ... The more scattered the population, the better for the insurgent ... A high ratio of rural to urban population gives an advantage to the insurgent.
8. *Economy* ... A highly developed country is very vulnerable to a short and intense wave of terrorism ... An underdeveloped country is less vulnerable to terrorism but much more open to guerrilla warfare, if only because the counterinsurgent cannot count on a good network of transport and communication facilities and because the population is more autarchic.

To sum up, the ideal situation for the insurgent would be a large land-locked country shaped like a blunt-tipped star, with jungle-covered mountains along the borders and scattered swamps in the plains, in a temperate zone with a large and dispersed rural population and a primitive economy.[6]

If Galula's experience had included Latin America, he might well have written that the ideal country for the insurgent would be Colombia, since conditions there in 1964 (and for many decades thereafter) were favourable to the insurgents in seven of the eight factors he lists. In the five decades since Galula's book appeared (a period that happens to coincide exactly with the evolution of FARC), the guerrilla movement rose from fewer than fifty fighters in 1964 to more than 20,000 in 2002 (before falling to roughly 6,900 today), becoming Latin America's oldest and largest insurgency along the way. The conditions Galula identified in 1964 continue to influence guerrilla and counter-guerrilla warfare in the Colombian conflict, and it is thus worth examining each in turn.

2. Colombia: Terrain Versus Population Density

(a) Terrain

2. Colombia: Terrain Versus Population Density

(b) Population Density

As can be seen from the maps, Colombia lacks significant natural barriers—its southern and eastern plains merge into those of northern Peru, northwestern Brazil and western Venezuela; its Andean frontier and coastal strip in the south-west blend into northern Ecuador; while its border with Panama is political rather than geographic. The Caribbean Sea to the north, like the Pacific Ocean to the west, is actually not a barrier, but rather a maritime highway for coastal and trans-oceanic shipping traffic, intra-regional and international trade, and a host of licit and illicit economic activities that connect Colombia with its neighbours. Likewise, Colombia's neighbours have historically been neutral or favourably disposed towards Colombian guerrilla movements—including FARC—or have been unable to control their border regions, creating space for guerrilla safe havens astride Colombia's frontiers.[7]

Colombia is very large—the fourth-largest country in South America (with a land area of more than a million square kilometres, and 3,000 kilometres of coastline on the Caribbean and the Pacific)—and, with 46.2 million people, is second only to Brazil in population.[8] As the population map shows, however, human settlement is extremely heavily concentrated in the western third of the country, which is made up of a narrow coastal strip along the Caribbean and Pacific coastlines, a highland plateau and the extraordinarily rugged *sierras* of the Eastern, Central and Western Cordilleras. These ranges 'form the extended Magdalena and Cauca river valleys [west of the dotted line on the maps] where most of the population resides. Because of the difficult terrain, most people live in fourteen main clusters of "city-states", each with a distinct economy and social character.'[9]

This uneven population distribution relative to terrain makes Colombia extraordinarily difficult to govern. It creates a diverse set of operating environments, exposes a sparse rural population to insurgents in the country's periphery, and generates densely populated but poorly integrated urban areas that are vulnerable to crime and terrorism and hard to police. Colombia's mountainous jungles—the third most rugged terrain of any country on the planet[10]—also make the country extraordinarily difficult to compartmentalise, letting guerrillas move freely and create safe havens undisturbed by the state.

Out of all of Galula's insurgent success factors, only Colombia's climate—equatorial jungles, crocodile-infested swamps, high-altitude mountain peaks, some of the wettest places on earth, coastal and estuarine waters with up to a four-metre tidal range, and a vast inland riverine network—suggests significant difficulty for guerrillas, and this, of course, affects counter-guerrilla forces too.[11]

The territorial logic of stalemate

In part because of its history—Colombia was settled and developed from the centre out, rather than from the coastline in—and in part because of these geographic and demographic conditions (which together create the 'territorial logic' of the conflict), Colombia in economic and political terms is really two countries, or rather one country embedded within another. Its economic and socio-political core is an urbanised, long-settled, relatively developed middle-income democracy (though also vulnerable to urban terrorism). This centre, however, is surrounded by an underdeveloped, unequal, historically ungovernable feudal-agrarian periphery lacking infrastructure and government presence, with a population that has historically experienced economic exclusion and political marginalisation, making it highly vulnerable to guerrillas who can exploit the legitimate (or, at the very least, completely understandable) grievances of poor rural *campesinos*.

The paradox of Colombia is that because of the country's uneven population distribution, even if every person in its rural periphery was an aggrieved, guerrilla-supporting *campesino*, they could only ever be a small minority of Colombia's overall population. This creates a built-in dynamic of stalemate in guerrilla warfare: because the agrarian poor have historically been a small minority, out of sight and out of mind for most Colombians, a series of democratically elected governments in Bogotá since the nineteenth century has lacked strong incentives to address peasant grievances, making uprisings and rural criminality more likely. At the same time, because marginalised *campesinos*, though a minority in the country overall, were often the majority of the population in any given rural district, guerrillas could expect significant popular support in these districts, allowing them to operate with impunity from a distant state, operating at the extreme edge of its reach, while hiding within the population like Mao's guerrilla fish in the sea.

Because disenfranchised and aggrieved peasants made up the bulk of the population in many remote districts in the country's marginalised periphery, Colombia experienced near-permanent banditry and guerrilla activity in these districts, making guerrillas extraordinarily hard to suppress as long as they stayed in the countryside. But in Colombia overall, such peasants were a small minority, concentrated in the remotest, least populated parts of the periphery. Thus the same guerrillas who could survive and thrive in the countryside could easily be defeated if they over-extended themselves beyond their rural sanctuaries to confront the state in the industrialised and urbanised centre of Colombia.

This territorial logic makes it virtually impossible for rural guerrillas to overthrow the Colombian state, yet also extremely difficult for any central government to stamp out a determined rural uprising, and this in turn has given rise to a longstanding tradition of low-level guerrilla warfare and banditry in Colombia—'a habit of "easy" rebellion against inept governments that were habitually inattentive to the needs of [the small minority of] their citizens' living in remote rural districts.'[12] In the twentieth century, urbanisation and population growth—the emergence of poor urban-dwellers, rural-to-urban migrants and urban fringe populations who could be exploited by urban guerrillas or criminal gangs—began to change this dynamic to some degree, tipping the balance against the state by the 1970s. In broad terms, however, the same conditions continued to exist.

The regional setting

But Colombia's internal circumstances are only part of the setting for its conflict, which is also heavily influenced by regional conditions. Colombia's hemispheric context is defined by competing influences: historical US economic and political dominance; competition from longstanding revolutionary Marxist movements in Cuba and Central America; the emergence of 'New Left' governments (in Bolivia, Ecuador and Venezuela) in the first decade of the twenty-first century; and the recent region-wide Bolivarian Continental Movement. External state actors, including the US, the Soviet Union (later Russia), China, and Iran, along with non-state actors including the IRA, ETA, Hezbollah, European socialist parties, and human rights groups, have shaped the environment for guerrilla and counter-guerrilla warfare in Colombia. The influence of this regional system can be seen most clearly in the interplay of narcotics and insurgency in Colombia.

The rise of a regional drug economy in the 1970s, responding to competing demands from US consumers seeking cocaine and US governments determined to deny it to them, gave Colombian guerrillas access to vast resources—so vast, in fact, that the very scale and availability of these resources transformed FARC and other Latin American groups into self-funding conflict entrepreneurs (see Introduction). Over time, they became so unlike classical insurgents that the new terms 'narco-terrorist' and 'narco-guerrilla' had to be invented in the 1980s to describe them.[13]

US dominance of high-value commodity markets was nothing new in the western hemisphere—companies like United Fruit (parent to today's

Chiquita brand) had achieved huge political and economic influence in Colombia, Ecuador and the West Indies by the first half of the twentieth century through licit exports of bananas, while US oil and mining companies dominated Colombia's extractive sector. The frequent use of American military power and diplomatic influence to preserve the commercial dominance of American companies led Major General Smedley Butler, one of the most highly decorated US Marine Corps officers of the century (who served in the 'Banana Wars' in the Caribbean), to describe war as a racket.[14]

But the drug trade was different. The biggest new factor was the vast amount of money involved—in part because of the immense cash value of narcotics, and in part because competing players in the United States funded both sides of the drug war, with drug consumers funding the narcos and government funding the counternarcotics effort, thus effectively doubling US investment in the cocaine economy. The violence inherent in illicit markets gave guerrillas—and later paramilitaries—a comparative advantage that quickly gained them a dominant position. As James Henderson points out,

> Illegal drug money has been the common denominator of Colombian violence since the 1970s. Abundant cash generated by illegal drug sales fuelled the traffickers' violence against one another, and against the Colombian state and its citizenry. Colombia's guerrillas ceased being innocuous only when illegal drug money started cascading into the country. Their successes from the 1970s onward were, ironically, a function of the most savage form of capitalism: the illegal consumer-producer market dynamic.[15]

As explained in Chapter 1, FARC initially taxed coca growers and subcontracted to, collaborated with, and extorted funds from drug traffickers. Later, after the collapse of the Medellin and Cali cartels, the guerrillas took over large parts of the narcotics process, produced and distributed cocaine on their own account, excluded other players from the drug trade, and, in places like the San Vicente del Caguán demilitarised zone (DMZ) in 1998–2002, constructed entire political economies around narcotics.[16] This ready access to drug money had four effects on the guerrillas' operational system: it reduced their dependence on external sponsors, made them less reliant on popular support, allowed them to develop a vast overseas financial hinterland, and tied them to fixed positions.

Whereas 'FARC survived in the 1970s thanks to support from poor *campesinos* and external ideological partners, such as Russian and Cuban communist groups', once the drug trade took off in the second half of the 1970s this dependence was sharply reduced.[17] Access to independent sources of funding from narcotics, kidnapping and extortion made FARC less reliant

on communist bloc sponsorship, letting it develop a heterodox political platform outside the general communist line from the 1970s onward. This
autonomy also allowed FARC to develop its own theory of guerrilla warfare,
melding Mao's Protracted People's War with Guevara's *foquismo*, after its
Seventh Conference of 1982. FARC's autonomy alienated it from official
communist parties in Colombia and elsewhere, encouraged its independent
streak vis-à-vis Cuba's Castroist regime and other left-wing governments in
the region, and helped it survive relatively unscathed when the Soviet Union
collapsed in 1991.

Secondly, funding from narco-traffickers and, later, their own drug sales
made Colombian guerrillas less dependent on popular support. Classically,
insurgents depend on local populations for recruits, intelligence, money, food
and shelter, while drawing on adversaries (government forces or rival guerrillas) or external sponsors for weapons and military supplies. The classical guerrilla centre of gravity—the element from which they derive their strength and
freedom of action—is their ability to manipulate and mobilise support from
a mass population base. This in turn makes preserving access to the population
and creating supportive (albeit often exploitative) relationships with communities a critical guerrilla task. In Colombia, this had been the traditional
guerrilla approach for centuries. But for the FARC in the 1980s, the new ease
of access to drug money supplanted access to the population as the critical
source of support, while narcotics networks supplanted popular mobilisation
as the centre of gravity. To the extent that guerrillas were tied to protecting
and supporting the interests of any population group, that group became the
small subset of Colombia's rural population who were actively engaged in drug
cultivation and processing, rather than the people as a whole. This limited
FARC's appeal to a subset of a subset within the wider Colombian population, perhaps 5 per cent of the population at most, and ultimately imposed
very serious constraints on its ability to expand its influence.

At the same time, FARC used drug funds to purchase weapons on the international market, finance propaganda and political activities, and sustain guerrilla structures. This lack of a need to generate sustained and widespread
popular support made the guerrillas less careful about civilian casualties—as in
May 2002, when one of FARC's notoriously inaccurate *rampla* gas-cylinder
mortars killed 112 women and children in the church of Bojayá, Chocó during
a battle to control a lucrative drug- and weapon-smuggling corridor.[18] FARC
could get away with this disregard for civilian life—along with other practices
that strongly alienated the population, including kidnapping, extortion, and

forced recruitment of child soldiers—because it depended on criminal revenue to sustain its system, rather than relying on popular mobilisation.

Third, narcotics internationalised FARC's interests, driving the guerrillas to develop an overseas financial structure estimated in 2012 to account for more than 70 per cent of the group's assets, and which includes holdings in Venezuela, Ecuador, Costa Rica, Panama and Mexico—as well in Norway, Holland, Denmark, Germany, and Sweden.[19] Given annual drug revenue of roughly $1.1 billion, this created a vast financial hinterland for the guerrillas, letting them use overseas funds to recover from losses within Colombia, or continue the conflict after (or in parallel with) peace talks. The fragmented nature of the financial structure—each FARC front manages its own finances independently, while many assets are held by the international commission (COMINTER)—makes this part of the guerrilla system both resilient to disruption and difficult for FARC's own leaders to control.[20] Indeed, the sharp increase in coca cultivation and export in 2014–15—an increase that has taken place against the background of peace talks, the suspension of aerial spraying, and a reduction in military activity—emphasises how extraordinarily resilient this element of the FARC system remains, even as it calls into question the long-term effectiveness of counter-drug operations under Plan Colombia.[21]

Finally, as Fidel Castro argued in his 2008 critique of FARC (*La Paz en Colombia*), the guerrillas' dependence on drug trafficking—and on the associated kidnapping and extortion revenue—tied them to fixed locations (coca fields, drug labs, airfields, trafficking routes), making their movements predictable and helping security forces find and fix them.[22] Likewise, in the 1990s, reliance on criminal revenues pushed FARC into premature expansion to increase its coca holdings, brought it into conflict with drug cartels, and exposed it to military interdiction. Later, the need to guard captives and control captured towns soaked up resources—a 2002 analysis suggested that 'FARC devotes 37 fronts, some 2,800 men (50 per cent of its force), to drug-trafficking activities'—and effectively pinned guerrilla forces down as a garrison protecting their narcotics holdings.[23] While it is easy to discount Castro's criticism given the idiosyncrasies of Cuba's own revolution, it is certainly true that holding fixed locations is something guerrillas typically avoid at almost any cost, since it is so often the prelude to their destruction.

The guerrilla operational system

As already discussed, it was against this domestic and regional setting, at its Seventh Conference in 1982, that FARC evolved its operational system founded on 'the combination of all forms of struggle' and a New Mode of Operations (Nueva Forma de Operar, NFO). The combination of all forms of struggle integrated social, political, economic, ideological, diplomatic, and guerrilla campaigns; the emphasis would shift from one form of struggle to another in response to changing circumstances. This strategy, discussed further in Chapter 5, created a flexible approach that helped FARC adapt to military reversals after 2002. The guerrilla system that emerged from the NFO includes five main elements: support networks, military structures (including the main force, urban militias and terrorist cells), clandestine political movements, the FARC central structure, and the COMINTER. Each of these elements is worth examining as part of the overall FARC system.

Support networks

FARC support networks are made up of an underground comprising sympathisers and supporters, along with an auxiliary or supporting logistic network. As clandestine organisations, the size of FARC's auxiliary and underground is hard to calculate, but 2014 estimates suggest it may number as many as 22,000 members. Network organisation often takes the form of clandestine cells located in (or on the edge of) urban districts or villages. Members of FARC support networks blend into local communities and—wherever possible—avoid interaction with security forces, though their identities are often known to at least some members of the civil population in the areas where they operate. Underground members conduct reconnaissance, collect intelligence, engage in propaganda, pass messages among guerrilla units, and inform on members of the local population and the security forces. The auxiliary helps the guerrilla forces store and transport weapons, maintains ammunition and explosives caches, gathers and transports food and medical supplies, and assists in money laundering and black-market currency exchange. Members of support networks often have longstanding relationships with specific FARC leaders and guerrilla units, but may be motivated by commercial, family, or criminal—rather than purely political—interests.

Military structure

As of early 2015, FARC's military structure comprised roughly 7,000 combatants—down from almost 20,000 in 2002—and was organised into three components: the main force, militias, and terrorist cells. The main force—organised since 1982 in a hierarchical military structure, the people's army (Ejercito del

Pueblo, EP)—is based in remote rural areas and composed of full-time guerrillas. It comprises five blocs and two combined joint commands, sixty-eight numbered fronts, nine named fronts, seventeen mobile columns and thirty-three independent companies. The blocs and joint commands are responsible for a given geographical area, and bloc headquarters control both territorial and mobile forces within their zone. The mobile and territorial forces within each region are organised into fronts—combat units of varying size and function—that are responsible for planning and conducting their own operations, supporting themselves financially, and generating funding for their parent bloc and the FARC central structure. Under pressure from security forces, only about half of the fronts within FARC's military structure remain operationally active as of 2015, though those fronts that do survive have shown impressive capacity for regeneration, replacing as much as 90 per cent of their losses over the course of the campaign since 2002. Several independent fronts exist, representing elite mobile forces attached to bloc headquarters; most numbered fronts are territorial units that operate in a single zone, control a specific population, target a specific aspect of the industrial or government system (for example, an oil pipeline or a port), or protect an element of FARC's narco-trafficking enterprise; named fronts often operate in a more mobile manner. Fronts are organised into columns—the basic fighting unit of between twenty and forty guerrillas—with larger, more capable mobile columns conducting offensive or high-value tasks. Independent companies operate like mobile columns, but are smaller and often more specialised.

Members of main-force guerrilla columns are full-time members of FARC, recruited from rural and urban populations and from other insurgent and criminal groups. They wear military uniforms and insignia, hold formal military rank, carry both small arms and crew-served weapons, and include specialist heavy-weapons teams, snipers, IED specialists, communications teams and reconnaissance units. For a period in the late 1990s and early 2000s, the level of equipment and weaponry fielded by FARC main force guerrillas was at least as good as, if not better than, that of the Colombian army and police units they were likely to encounter in their operating areas. Since then, under pressure of military operations, all FARC columns—whether territorial or mobile columns—have been forced to become more mobile and to move back from the edges of urban areas. At the same time some elite groups, such as the Teofilo Ferrero Mobile Column, have been severely damaged and forced to curtail their operations.

The militias are a looser, cell-based, element that has become increasingly important since 2011 as pressure from the Colombian army's Joint Task

Forces (JTFs) destroyed much of the main-force guerrillas' military capability and forced them to withdraw to jungle bases or curtail offensive operations. Urban militias are organised into four named fronts targeting Colombia's major cities. They may be linked to specific blocs, or form part of the clandestine political structure (discussed in the next section). Urban militias operate in small groups up to a maximum of about twelve people, and maintain safe houses in urban or peri-urban areas, along with independent escape and infiltration routes. Each militia front is led and supported by a small cell of full-time cadres who can draw on a larger pool of part-time members within an operational area. The militias operate in civilian clothing with light, concealable weapons and rely heavily on explosives—though heavy weapons such as gas-cylinder mortars or heavy sniper rifles may be used for specific high-value operations. For survivability, members of urban militia fronts tend to live in one district, hold cover jobs in another, and operate in a third district within the same city or even in a different town. Weapons and explosives are moved and cached by specialists within the fronts, and may be kept separate from cell members until just before an attack, allowing 'clean' infiltration by an assault team into a target area.

Militias (also known as Bolivarian militias) operate in both rural and urban areas. Rural militias operate as part-time auxiliaries in support of a Main Force front and traditionally depended on that front. They provide intelligence and logistics for Main Force units and execute terrorist attacks in rural and urban environments. Their role has evolved, however, as the Main Force has been weakened through the COIN campaign since 2002. As a result, more rural militias have received formal military training and it's increasingly common for militias to operate autonomously.

Urban militia operations typically involve asymmetric tactics such as bombings, kidnappings, sabotage, blackmail, assassinations, racketeering, support to drug trafficking, and operations in conjunction with FARC's criminal alliances. The urban militias have successfully adopted tactics and weapons from international terrorist groups including IRA and ETA (including remotely-fired mortars, IRA-designed *ramplas*, Hamas-designed Qassam rockets, and several IED designs) and may have modelled their cell structure on the IRA's Active Service Unit system. At their peak in 2003, the urban militias included as many as 12,000 combatants, focused on Bogotá and Medellín but also targeting mid-sized cities like Bucaramanga. Under military and police pressure, these numbers have been reduced, and FARC urban tactics have shifted towards reconnaissance and intelligence—but the primary role of the urban militias remains asymmetric combat within Colombia's cities.

Similar to militias in their operational methods, but differing in terms of command and control, terrorist commands are small independent cell structures that are tasked directly by FARC central command (or by a front commander) for specific clandestine operations—bombings, high-profile murders and kidnappings, or hijackings—knowledge of which is 'compartmented' to senior FARC leadership and thus may not be known to, or coordinated with, local guerrilla commanders. These terrorist commands primarily operate within Colombia, but have the capability to operate internationally.

Clandestine political structures

Clandestine political structures have existed alongside FARC's military structures since its earliest days, but have evolved over time. They currently include the Partido Comunista Clandestino Colombiano (Clandestine Colombian Communist Party, PC3) and the Movimiento Bolivariano por la Nueva Colombia (Bolivarian Movement for a New Colombia, MBNC), both founded in 2000. The PC3 was led by Guillermo León Sáenz Vargas (alias 'Alfonso Cano') until his death in 2011, and is structured as an illegal, underground political party that controls the more overt MBNC, sponsors attacks by FARC urban militias, engages in agitation and propaganda, subverts Colombian government structures and legal political parties, and manipulates social movements such as trade unions and indigenous communities. The PC3 emerged after FARC's split with the official Colombian Communist Party in the early 1990s—an example of FARC's ability to operate outside the formal communist line due to its independence through access to drug money. The PC3's current approach—a clandestine, cell-based party controlling broader, more overt political movements—is a classic communist subversive technique, but it also represents an adaptation to the bloody and disastrous failure of the FARC-sponsored Unión Patriótica (Patriotic Union, UP), which as we saw in Chapter 1 was an overt, legal popular front party that drew broadly from Colombia's left, and was savagely suppressed through assassinations and massacres by narco-traffickers, paramilitaries and rogue members of the security forces in the 1980s. PC3, because of its clandestine subversive nature and its control of FARC urban militia fronts, can be considered a hybrid structure—both a military and a political element of the guerrilla operational system—that serves as the connective tissue between FARC's 'armed struggle' and its political and ideological campaigns.

FARC central structures

These military and political components support, and are (at least theoretically) controlled by FARC's central structures, including its General Staff and Secretariat. The Central Headquarters (Estado Mayor Central) is responsible for appointing, controlling and supporting the commanders of each bloc, delineating their areas of operation, collecting funds from blocs, resolving disputes between them, organising joint operations among columns from different blocs, and managing centralised functions such as intelligence, training (via the FARC military academy) and specialised weapons, intelligence and explosives cadres. This sounds like a busy job but, in practice, FARC blocs (as well as the combined joint commands, and often fronts themselves within blocs) are operationally autonomous—increasingly so, as the military high-value-target campaign has killed senior FARC leaders and forced decentralisation since 2007.

Within the central headquarters, the FARC Secretariat (Secretariado del Estado Mayor Central) is a small decision-making body—with only seven members for much of its history—that sets overall strategy, coordinates propaganda and political activities, and acts as the core command group for the whole movement. Since 2002, six Secretariat members have been killed or died of natural causes (possibly aggravated by the stress of pursuit through Colombia's mountainous jungle). Those killed included Alfonso Cano, Mono Jojoy, Raúl Reyes and Iván Ríos, while FARC spokesmen claim that Manuel Marulanda (alias 'Tirofijo') and Efraín Guzmán died of natural causes.

Further, since the start of peace talks in 2012, FARC leaders have taken the opportunity to join the talks and escape the increasingly dangerous field environment in Colombia—as of late 2014, the Havana group included three out of seven members of the Secretariat, three out of six known bloc commanders, and four sub-commanders.[24] This exodus of senior talent from the field has created a drain on FARC's leadership within Colombia, leading to further operational decentralisation, with many FARC fronts now operating relatively independently.

COMINTER

The international commission (COMINTER) is FARC's network of overseas representatives, sympathisers and supporters. Headed by Raúl Reyes until his death in 2008, the COMINTER includes FARC representatives and front organisations in twenty-seven countries. It functions as an international political, financial and subversive support network for FARC. As explained in Chapter 1, according to documents captured in the Reyes raid in 2008, the COMINTER's international strategy had four key objectives: to acquire financial and military support from overseas state and non-state actors, including advanced weapons systems such as man-portable air defence systems; to build political legitimacy (including recognition of belligerent status) for FARC while reducing international support for the Colombian government; to undermine Colombia's security cooperation with neighbouring countries; and to establish cross-border safe havens in Venezuela, Ecuador and elsewhere.[25]

Operational methodology

Within this relatively stable operational structure, the operating method of FARC has changed significantly since the 1990s, as discussed in the Preface and in Chapter 1. With the adoption of the NFO in 1993, FARC began a build-up to a semi-conventional military force. It created officer and NCO academies, centralised unit structures, created specialist units, adopted formal military organisations, ranks and uniforms, and attempted to turn itself into something resembling a regular army—as indicated by the addition of 'Ejercito del Pueblo' to its name in 1993. In part, this was an attempt to meet the requirements under international law for recognition as a legitimate combatant in an intra-state conflict, as a way of bolstering FARC's credentials as a counter-government.

In part, also, this transition to conventional conflict was a mark of FARC's confidence in its strategy of encircling the cities by occupying and controlling peri-urban terrain and smaller towns and routes between Colombia's major urban areas. The model for this approach was the successful Sandinista movement in El Salvador, which FARC cadres studied in detail in their academies, and which led FARC to transition to a pseudo-conventional 'war of movement' in the 1990s—only to be forced to drop back a stage to guerrilla warfare (with the addition of a significant upsurge in urban militia terrorism) after the death of Raúl Reyes in 2008 and the adoption of Alfonso Cano's Plan Renacer.

As of mid-2015, the strategic purpose of FARC's remaining main-force guerrilla units, in addition to rural guerrilla operations, appears to be twofold: to maintain a force-in-being allowing FARC to claim legitimate combatant status and bolster the credibility of FARC negotiators in Havana, and to preserve and protect FARC's bases and illicit sources of revenue. This represents a shift from the late 1990s and early 2000s, when FARC sought to encircle and conquer cities in Colombia's industrialised core. This strategy,

> ... envisioned surrounding the national capital with a force of 16,000 armed fighters, cutting the city's food supply, and thereby causing a general uprising against the government. When that came to pass FARC leaders would emulate both Fidel Castro and Daniel Ortega and ride triumphantly into Bogotá at the head of their rebel army. The NFO was fanciful when announced [in 1982] as the FARC's total armed force consisted of just 1,000 men. Yet thanks to the peace initiative of Belisario Betancur the FARC grew virtually unimpeded for four years, more than tripling by 1986, to 3,600 fighters. The encirclement and seizure of Bogotá was part of the FARC master plan for eventual triumph drawn up at the 1982 meeting. Called the Strategic Political-Military Plan (*Plan Estratégico Político Militar*), it projected continuation of the insurgents' early 'centrifugal strategy' of continuously spinning new fronts out into key parts of the nation. The plan was spectacularly successful, for it allowed the FARC to increase its fronts from seventeen to thirty by Betancur's last year in office.[26]

As noted earlier, this attempt to fight the government on its own turf breached the system boundaries of Colombia's longstanding territorial logic of guerrilla stalemate, prompting a massive countervailing response. When the Colombian government pushed back in 1999–2002, recapturing control of cities and surrounding districts, FARC was forced to drop back to a guerrilla approach, and then, following losses of key FARC commanders in 2008, to what might be described as an 'enclave' strategy. This combined a rural guerrilla force-in-being—which fulfilled its purpose simply by continuing to exist and generate revenue—with increased urban guerrilla warfare and terrorism by the urban militia fronts. Under Alfonso Cano, Plan Renacer sought to disperse fronts and columns into smaller guerrilla units to increase their survivability against air strikes and army operations; to move in a dispersed formation of small groups rather than in large conventional battalions; to operate in more remote and rural terrain; to increase engagement with the population in order to rebuild the popular support FARC had squandered in its highly aggressive mobile warfare phase; and to strengthen the urban militia fronts. Despite significant losses since 2011, this remains the basic FARC structure and operational posture in 2015.

Issues in Colombia's counter-guerrilla system

As discussed in Chapter 1, and as this account of FARC's forced adaptation in the face of the Colombian military's counter-guerrilla operations since

2002 demonstrates, Colombia has made huge progress in reducing the threat FARC posed to the country's urbanised and industrialised core. Yet, despite a turnaround so dramatic that some analysts have called it 'The Colombian Miracle', it should be clear that Colombia still faces a robust insurgency.[27] Four specific issues, as of 2015, include: long-term sustainability; the inability of civil agencies to build effective COIN programmes to match the successful counter-guerrilla warfare efforts of the military; the problem of village governance and security; and FARC's criminal alliances (perhaps better described as the convergence between crime and guerrilla warfare).

The first critical issue is sustainability. In driving the guerrillas away from their attempted encirclement of its cities, the Colombian government relieved the national crisis of 1999–2002, but the danger now is that, having re-established the traditional equilibrium—the state controlling the centre, the guerrillas maintaining safe havens in the country's periphery—the dynamic of stalemate will reassert itself and Colombia will fail to take the necessary steps to resolve the conflict and prevent its resurgence. What Colombia needs now is a sustained counter-guerrilla effort—perhaps lasting between fifteen and twenty years—to consolidate the gains achieved since 2002. The risk is that these gains could potentially be undone overnight in the event of failed peace talks, an economic setback that forces cuts to the defence budget or a reduction in the number of troops deployed, or a lopsided peace agreement that allows FARC to recover and build back its strength.

More fundamentally, there is an asymmetry of objectives here: on the one hand, the government of Colombia seeks to end the conflict via a political settlement, (ideally on favourable terms) with the guerrillas. On the other, many senior and mid-level FARC commanders, as conflict entrepreneurs, may seek not to end the conflict, but to preserve it so as to maintain access to ready sources of drug funding. Despite huge progress against the guerrillas since 2002, Colombia's cocaine production and smuggling networks remain extremely vibrant, and as long as the guerrillas can access this source of funds they can sustain their activities more or less indefinitely, even in the absence of significant popular support. In any case, under the strategy of the combination of all forms of struggle, FARC leaders appear to regard peace talks as just one more phase in an ongoing struggle that serves their business interests as much as their political goals.

For its part, Colombia's military, after its hard-won battlefield successes, a huge expansion in its numbers and combat capability (as discussed in Chapter 2), and a massive growth in public support and prestige, faces a different dilemma. Military commanders understand they must sustain a local security presence, and remain involved for the foreseeable future in governance and economics via programmes such as Integral Action (local-level reconstruction and community engagement projects sponsored by the military), in order to enable civilian agencies to work with the population, extend governance, improve services, and reduce the structural inequality and exclusion that provoked the insurgency in the first place. This effort will take enormous political commitment over a long time—particularly since historical benchmarks suggest that post-conflict stabilisation may take twice as long as the conflict that preceded it.[28] But such a commitment—on top of the massive growth in military budget, manpower, and prestige of the past decade—brings personal and institutional incentives that carry the risk that the military, too, may become stakeholders in a political economy of war, with institutional interests in preserving the conflict. This risk may be worth taking—without a sustained counter-guerrilla presence in contested areas, it is hard to see how the conflict can end—but it is still a risk.

A second critical issue lies in the emerging gap between the military's efforts in counter-guerrilla warfare, and the broader whole-of-government COIN effort that (in theory) should surround and complement the efforts of security forces. As we have seen, military counter-guerrilla efforts have rolled FARC back into a defensive posture, but civilian agencies need to step more actively into the space created by these military operations, lest soldiers be left holding an empty bag—or, worse, become tempted to usurp civil authority in order to get the job done.

As originally designed, the Sword of Honour campaign plan envisaged civilian agencies assuming administrative functions in contested areas and police taking over cleared areas once the military had defeated or displaced the FARC main force. This was intended to free military forces (led by the thirteen mobile JTFs established under Sword of Honour) from a ground-holding role, allowing them to manoeuvre against FARC base areas. Under Green Heart, the National Police companion plan to Sword of Honour, police were to assume responsibility for protecting cleared ('green') areas and for dealing with FARC auxiliary and underground networks, freeing the military for mobile operations in contested ('amber') areas and FARC base ('red') zones. In practice, rural populations—who had a long history of negative interactions with the police—rejected police presence in these areas, preferring that the military remain in place. At the same time, civilian agencies of government proved unable to fulfil their role in a timely and effective manner.

Thus, as noted in Chapter 1, the military was forced to expand its activities beyond counter-guerrilla warfare to take on broader civil COIN functions that could properly be considered the role of other agencies of government, leaving troops pinned down in administrative, governance and Integral Action roles, rather than doing what only soldiers can do—keeping the enemy under pressure, in order to set the military conditions for successful peace talks to resolve the conflict. As one analyst argues, Colombia is not 'killing FARC fast enough to put enough pressure on them to achieve a peace settlement, because we're soaking up the army doing things that are really the job of civilians'.[29]

Thus, while the counter-guerrilla system built by the military is impressive, the complementary civilian programmes needed for a true, full-spectrum COIN system are lagging, creating gaps that the guerrillas can exploit. Unless the government creates a permanent presence at village level to replace the rival FARC guerrilla system that has dominated communities for so long, destroying today's guerrillas will only create a vacuum to be filled by successive generations of insurgents and criminals. This is a fundamental challenge: counter-guerrilla progress without comparable improvements in civil governance either leaves the military pinned down, protecting every bridge, schoolhouse and government office—or it makes the population vulnerable to guerrillas once the military leaves.

The third key issue is that of village-level governance and security. However effective its mobile counter-guerrilla operations (carried out by the JTFs) or its local security operations (carried out by territorial units) may be, the Colombian military does not in fact maintain a permanent presence at village level—troops establish bases at *municipio* level (the equivalent of a US county or British local authority), or in departmental capital cities, and then send out patrols that visit villages only periodically, and stay only briefly. To avoid violating civilian property rights by sleeping in schools or private houses—a practice the army banned several years ago as part of its enhanced focus on legitimacy through human rights—patrols rarely sleep in villages but instead retire to patrol bases in the jungle at night.

This means that, when a patrol arrives in a village, guerrilla forces simply retreat to the hills or nearby jungle. The underground cells and members of the FARC auxiliary (described earlier in this chapter) remain in place, however, maintaining a network of informers to monitor the villagers' behaviour while the guerrillas are gone. As soon as the soldiers leave, the guerrillas return. Villagers who interact with soldiers on patrol in rural areas therefore know that as soon as the sun goes down, or at most within a few days, the guerrillas will hold them to account for whatever they do or say.

Civil government also lacks a permanent presence at the village level: the mayor of each *municipio* (which, in some districts, may include more than 200 individual villages) represents the lowest level of formal administration within the Colombian government structure. Most mayors remain in their offices in district capitals, rarely visiting villages and instead interacting with village-level Community Action Committees (Juntas de Acción Comunal, JACs), informal bodies with no status under the Colombian constitution. Most governance at village level—and thus, by definition, most of the governance that matters in the sense that it directly affects the rural agricultural population—is performed by JACs, and where FARC has a permanent presence, the guerrillas influence membership in the JAC, applying their concept of 'co-government' (described in Chapter 1) to manipulate and control local governance structures so that, in the words of one soldier, in some areas 'the neighbourhood council *is* the guerrilla front'.[30]

This dynamic—fragmented or absent civil governance and episodic military presence, combined with permanent local presence of the guerrillas—creates a double 'brain drain' at the village level. Government supporters are systematically culled as the military's episodic visits expose them to retaliation from the guerrillas, but then leave them unprotected. Meanwhile, villagers see FARC as the legitimate system, and those with talent and ambition disappear into the movement. Money, brains and jobs—in that order—flee contested areas, and once gone are extremely difficult to get back.

Indeed, it was the inability of the government to offer permanent protection to Colombia's rural population that gave rise to the paramilitaries in the 1990s. As noted in Chapter 1, the killing and intimidation of farmers, local small businessmen, prominent townspeople and ranchers led to the formation of self-defence forces—like Castaño's Autodefensas Campesinas de Córdoba y Urabá—which successfully drove FARC out of much of northern Colombia by the 1990s. But the rise of the *autodefensas* was in effect a vote of no confidence in the state's ability to protect the rural population, and led to severe human rights abuses, mass killings of FARC sympathisers and members of the UP, and ultimately the emergence of alliances between narco-traffickers, organised criminal gangs and the paramilitaries. The failure to maintain a permanent state-based counter-guerrilla presence in Colombia's periphery had simply enabled the rise of yet another illegal armed group.

This highlights a final key issue in the counter-guerrilla campaign: namely, the criminal alliances, or crime–guerrilla convergence, highlighted by the rise of BACRIM. FARC–BACRIM collaboration exemplifies the strange bedfellows that can emerge in a political economy of war, when conflict entrepreneurs see opportunities to perpetuate the violence from which they benefit. Most of the right-wing Autodefensas Unidas de Colombia (United Self-Defence Forces, AUC) paramilitaries were demobilised under the 2006 agreement brokered by President Uribe—but by that time some had become little more than bandits, and these groups evolved into paramilitary criminal gangs, BACRIM.

Having given up their anti-FARC political agenda after 2006—it was now all just business—BACRIM were willing to collaborate with anyone (including FARC) who could advance their goals of plunder and profit. For its part, once FARC came under sustained counter-guerrilla pressure, insurgent leaders saw an opportunity to spread government efforts more thinly (as part of the 'centrifugal strategy' described earlier) by establishing temporary alliances of convenience with criminal groups, using proxies to protect their cocaine economy, and hiding within criminal networks. This was a sensible strategy for FARC: a conflict entrepreneur, with commercial rather than political interests (as was the case for many FARC leaders by 2006, given the long-standing relationship between the insurgent group and the drug economy) has nothing to gain by posing as a politically motivated insurgent. On the contrary, given the effective HVT campaign conducted by Colombian security forces, this would run the risk of being killed in a special-forces raid or targeted in an air strike. But by adopting the identity of a criminal, such an entrepreneur can make more money, and have a more comfortable life, protected by Colombia's robust civil-rights and criminal-law protections, while risking nothing more serious than arrest in a country with no death penalty.

Dealing with this kind of ingrained violent criminality demands more than police: a viable justice system must include courts, corrections and effective formal and informal legal institutions. But delays in the court system, and overcrowding of jails, mean that some detainees end up serving their sentences in holding cells in police stations. Judges shy away from custodial sentences for all but the most extreme crimes, realising that there is no room for more detainees within the correctional system—hence many violent offenders, even notorious BACRIM and guerrilla members, are quickly released. This in turn frustrates police and military officers, who see known criminals and insurgents walking free, able to retaliate against witnesses. Over the long term, such impunity—for individuals who the community and law enforcement are convinced are guilty—can prompt ordinary people to take the law into their own hands and (in extreme cases) result in extrajudicial killings. But change is hard to imagine without a structural shift in the incentives that turn people into conflict entrepreneurs—in other words, without transformation from a political economy of crime and conflict, to one of sustainable and inclusive prosperity and peace. This is a daunting challenge, but it is the fundamental task of peacebuilding: after generations of conflict, it should be no surprise that making peace should be difficult or require a wholesale transformation.

Insights

From this account of guerrilla and counter-guerrilla warfare in Colombia (and from the historical and capability discussion of previous chapters), it seems clear that the conflict is at a complex inflection point. The insurgency is far from spent: a sizeable force of guerrillas remains in the field, and even many demobilised fighters remain committed to revolutionary ideologies, and might vote for MBNC candidates or even for overtly FARC candidates were the guerrillas to again create a legal political party. The prospect of peace offers FARC, within its 'combination of all forms of struggle', the opportunity to trade a tenuous military position for a stronger political one through negotiation. It may seek to manipulate grievances, mobilise populations and capture the state through the ballot box, a 'revolutionary judo' move like Bolivarian revolutionaries elsewhere. Moreover, peace offers FARC racketeers the option of dropping their political agenda and (like the paramilitaries before them) redefining themselves as BACRIM. This suggests several insights for Colombia's future, and for other countries experiencing similar challenges.

First, in political terms, the irony is that the military's very success in counter-guerrilla warfare may undermine support for its ongoing stabilisation efforts in contested areas. As we have seen, the terrain and population distribution of Colombia create a built-in tendency towards stalemate in guerrilla conflict. Under their post-1993 mobile-warfare strategy, FARC—who had proven strong and resilient while remaining in Colombia's remote and rural periphery—overreached, attempting to encircle and capture the cities of the country's urbanised and industrialised core. This provoked a national crisis, and led to a mobilisation of state resources to relieve the threat to the cities. But as the military's counter-guerrilla operations successfully forced FARC back into the rural periphery and compelled it to drop back to guerrilla warfare, the perception of threat in Colombia's urban core (where, as we have seen, the vast majority of Colombian voters live) dropped—but so, as a result, did the public sense of urgency and thus the pressure on elected Colombian leaders to maintain stabilisation efforts. Many Colombians are ready for the war to be over, and now that FARC seems less threatening, other concerns predominate. This combination of war fatigue and shifting concerns on the part of urban Colombians helps the guerrillas: having pushed the FARC back into the rural periphery, there is now a risk that the state might lack sufficient popular support to finish the job, thus driving another cycle of the long-standing dynamic of guerrilla stalemate. The Colombian government's key COIN challenge is therefore to sustain political support without (on the one hand) letting voters slip back into apathy, or (on the other) putting Colombia onto a perpetual war footing.

Resolving the dynamic of stalemate will demand a genuine social transformation in Colombia's rural environments and among urban fringe populations—a transformation that fundamentally alters the terms of the conflict by creating a more inclusive society for the excluded and marginalised populations that are FARC's principal constituency. This in turn requires recognition that FARC and BACRIM are conflict entrepreneurs seeking to perpetuate violence for personal gain, so that extension of government presence and rule of law to the very local level of society is critical.

Related to this, given the current inability of civilian agencies to deliver the governance and reconstruction effects envisioned in Sword of Honour, political leaders need to recognise that the critical COIN element now, and into the future, is not the military's counter-guerrilla-warfare programme, but rather the ability of civilian agencies to complement this programme with a broader effort. Local civil governance—and the willingness of civilian agencies to support a comprehensive national plan—demands political leadership. Since civilian agencies are not under the authority of the defence minister, such leadership can only come from the presidency.

This suggests that a balance must be struck between pursuing peace talks on the one hand, and extending civilian governance and social programmes with sufficient energy and determination to free the military from static garrison and civil-affairs tasks on the other. Only then are the JTFs, territorial forces and strike assets of the military likely to generate enough counter-guerrilla pressure on the remaining elements of FARC's guerrilla system.

A third political insight is the recognition that the peace settlement for one conflict can sow the seeds of another. In the case of La Violencia, exclusion of some Colombians from the National Front government of 1958–74 contributed, in part, to the emergence of the 'independent republics'; the suppression of these republics created today's insurgency and turned rural self-defence groups into an active and widespread guerrilla army. A future peace settlement that lets conflict entrepreneurs unfairly control territory or monopolise government institutions could lead to a 'soft takeover' by groups that have been defeated militarily and are supported by very few Colombians.

But equally, excluding such actors from politics could set the conditions for yet another round of guerrilla warfare, and denying them economic opportunity could increase criminality, as insurgents rebrand themselves as BACRIM. Finally, a settlement that penalises members of the military or police for their actions during the conflict, while giving blanket amnesty to guerrillas, may create a constituency against political integration within the security forces—as soldiers and police worry whether some future government may punish them or their families for acts that were legal and seen as necessary at the time—with potentially dangerous effects on Colombia's civil democracy.

In parallel to these political observations, a key military insight is the requirement for counter-guerrilla forces to redouble their efforts to secure the at-risk rural population—those people who are willing to work with the government, but living in FARC-dominated areas. These people are the seed-corn of future rural stability, development and political progress, and must be protected at virtually all costs. Periodic raiding or patrol visits expose them to retaliation as soon as soldiers move on, and this in turn systematically culls community leaders in contested areas. The solution is likely to involve some form of enduring presence—in which troops live, permanently, among the people at village level, creating a safe enough environment that local communities feel confident to identify members of the guerrilla underground and auxiliary networks, and reversing the brain drain by helping JACs, community leaders, and talented local people regain control of their own villages. Where this is not possible, however, military commanders may want to think twice about overtly engaging with friendly members of the population—or at the very least, to give them some viable excuse by which to explain their interaction with the security forces once the troops leave.

In any case, effective protection of the local population is the first critical step to reverse the loss of rural money, brains and jobs by lifting the pall of fear from rural communities. A flexible counter-guerrilla system involving police or military outposts at village level, supported by district-level quick-reaction forces and embedded civilian administrative officials at the village level, could create a framework for radical improvement. Until civil agencies and police can effectively backfill the military, however, there will never be enough troops to secure all contested districts—and until that time, there will be a need to prioritise key districts and redirect effort away from tasks that are properly those of civilians, towards a single-minded focus on population security at the local level.

Over the long term, this counter-guerrilla effort (primarily the role of the army's territorial brigades) needs to be complemented by an effort to fully unleash the JTFs, accompanied by special forces and supported by air and maritime power, to radically increase pressure on remaining FARC structures and on BACRIM. The goal of such operations, in any campaign where they are undertaken, is not to kill or capture every last guerrilla, but to convince insurgent leaders (or negotiators in a peace process) that they are in a closing window of opportunity to achieve peace before their forces in the field are destroyed. In the crudest terms, a military engaged in counter-guerrilla warfare needs to seize control of the guerrillas' loss rate—driving that rate upward, until a sufficiently high rate of kills, captures and surrenders is achieved that adversary leaders understand their best option is a negotiated peace. At the same time, intensive targeting of organised crime groups by police (which must, needless to say, stay strictly within the bounds of the rule of law) can help convince insurgents that criminality offers no sanctuary.

One final issue concerns the future size, role and structure of Colombia's military forces, which, as noted, have grown significantly over the past decade. Typically, at the end of an internal conflict such as Colombia has experienced, disarmament, demobilisation and reintegration (DDR) of combatants—on all sides—is a crucial element both of peace talks and of post-conflict security sector reform. It may seem premature to consider possibilities for demobilisation and restructuring of Colombia's military—what we might call a 'peace dividend'—while peace is still in doubt. Likewise, to rapidly demobilise military forces without adequate alternative employment opportunities runs the risk of putting large numbers of young, capable Colombians out of work and on the street, where they might become vulnerable to retaliation by former guerrillas or recruitment by organised criminal groups. But political leaders are rightly concerned to ensure that Colombia, in the face of declining global oil prices, budget cuts and straitened economic circumstances, maintains a sustainable military, with appropriate roles and missions, after the end of the current conflict—not least, one that is able to prevent a resurgence of internal violence while furthering Colombia's regional interests and meeting its international obligations. In this context, structures like a national guard (which would allow demobilised veterans to serve part-time in their home districts) a rural constabulary (emphatically not a paramilitary organisation, but rather a part-time civil police auxiliary under regular police commanders, controlled by robust legal oversight and responsive to local civil authority), or a reconstruction corps that would provide employment and training to ex-soldiers and enable infrastructure development, are worth considering. Options like these would help create a pathway to peace that soldiers could understand and support, and prevent demobilised military personnel being drawn into criminality or destabilising political activity.

4

THE DOOR THROUGH WHICH MUCH FOLLOWS?

SECURITY AND COLOMBIA'S ECONOMIC TRANSFORMATION[1]

Greg Mills

The breeding ground for war is poverty and a lack of economic opportunity.
Hugo Morena, Councillor, La Macarena, Meta Department, June 2014

Having laid out in previous chapters the historical patterns of the conflict, the relationship between capability development and military success, and the dynamics of guerrilla and counter-guerrilla warfare in Colombia, this chapter examines the relationship between security and economic development in Colombia—the mechanism through which the dynamic of stalemate, identified in Chapter 3, will be broken, if indeed it can be broken at all. It is argued that security—and the resultant stability, predictability and business confidence—has been the door through which economic development in Colombia has followed. That development, in turn, is an essential step to ending the decades-long cycle of conflict in Colombia's marginalised and excluded periphery. This chapter also locates the role of international assistance in this period, and identifies both a number of future challenges and a range of ways in which they might be met.

The gondola of Medellín

The gondolas of Line J of Medellín's Metrocable sail high above La Comuna 13, one of the toughest *barrios* of the city. Inaugurated in 2007, using Swiss technology, the funicular connects the 28,000 members of the Comuna, among others, with the central business district (CBD) of Colombia's second-largest city. The journey, which would once have taken several hours of travel up and down winding, narrow roads, takes ten minutes, and costs just $1.

Looking down at the rusted tin roofs and redbrick dwellings perched on the hillside, a local policeman observed: 'We had a problem at the start of the cable-car. The locals were shooting at it from the ground'. This security problem now solved by improved patrolling, Line J, one of three spanning the city, carries 30,000 people a day, the gondolas travelling quickly over the *barrios* at 16 kilometres per hour, delivering their human loads efficiently at San Javier station at the bottom and at La Aurora on the top of the hillside 2.7 kilometres away.

Once in the CBD, commuters hop onto the Metro, first opened for service in 1995, and built by a Spanish–German consortium. Smart and litter-free, its twenty-seven stations and carriages are a metaphor for the contemporary change of Medellín's fortunes, from the town of Pablo Escobar to the epicentre of Colombia's mining and manufacturing industries. The Metro carries half a million passengers daily, including 350,000 residents from the north-eastern area where many of the working class live. In so doing, it is breaking down the barriers between once disparate poor and rich quarters, and enabling new business growth.

With construction costs for the Metrocable at $10 million per kilometre and the Metro itself costing $2 billion, it was a bold step. But there was comfort in historical precedent. The Ferrocarriles de Antioquia (Antioquia's Railways) had put the region in touch with Bogotá, Cali and the outside world, overseeing a period of intense industrial growth during the late nineteenth and early twentieth centuries. Rising gold and coffee exports followed, as did a growth in manufacturing.

The use of the Metro as a development axis was recognised by Medellín's planners as critical in meeting the city's modern needs in a period of social change and instability. For the pace of Medellín's urban growth since the

1960s had filled the entire Aburra Valley with communities, where harsh living conditions were heightened by drug trafficking, joblessness and violence. As noted in the previous chapter, the rapid pace of urbanisation from the middle of the twentieth century changed the traditional centre–periphery dynamic of Colombia by creating a growing, marginalised and excluded urban (or peri-urban) population. As a result of this—and the related rise of drug trafficking under kingpins like Pablo Escobar—Medellín boasted the highest rates of violent crime worldwide, with the rate touching nearly 7,000 murders per year at the peak of Escobar's reign in the early 1990s. By 2008, this was down to an annual rate of little more than 1,000 homicides, falling further by 2014 to 658.[2] In 1991, to use a different measure, Medellín experienced 381 homicides per 100,000 residents; twenty years later, in Ciudad Juárez, then the epicentre of Mexico's drug war, the rate was less than half that. Now Medellín has the same homicide rate as Washington DC.[3]

This turnaround has been described by Francis Fukuyama and Seth Colby as 'half a miracle'.[4] A lot of this success has however been down to better policy and assiduous application, and more investment in infrastructure. In a positive cycle, improved security has led to economic growth, which, in turn, has cemented stability. Medellín is Colombia's top export region, boasting some 1,750 export businesses, from textiles to services. This activity is supplemented by mining, electricity generation, construction, business and, increasingly, tourism.

Colombia's largest business conglomerate, Grupo Empresarial Antioqueño, which contributes some 7 per cent of the country's GDP, is also headquartered in the city. Like a Japanese *keiretsu* or Korean *chaebol*, it comprises four major

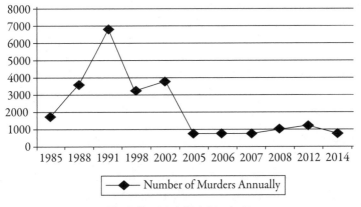

3. The Fall in Medellín's Murder Rate

elements: Bancolombia (banking), Inversiones Argos (the country's largest cement manufacturer), Suramericana de Inversiones or Inversura (South America's largest insurance group), and Grupo Nacional de Chocolates (packaged food). Antioquia, the department of which Medellín is the capital, is also the largest producer of gold in Colombia, officially producing 25 tons in 2015. This is likely to be driven up if some of Antioquia's exploration projects come to fruition. For example, AngloGold Ashanti's Gramalote Project, some 110 kilometres north-east of Medellín, has an anticipated yield of 8.7 million ounces (270 tons).

In good times and bad, therefore, the drivers of national growth receive a great deal of input from Medellín.

4. Sources of Colombia's GDP Growth, 2014–15[5]

Sector	2014	2015 (est.)
Agriculture	2.3	3.4
Mining (inc. oil)	−0.2	−1.1
Construction	9.9	6.9
Manufacturing	0.2	3.0
Other	5.3	4.1
Total GDP	4.6	3.5

The spending on Medellín's infrastructure and public spaces has helped not just to pump up demand but also to enfranchise those poorer neighbourhoods scattered over the hillsides, otherwise largely absent any public goods. This has helped to change local attitudes and integrate communities into urban life. Medellín gained the Urban Land Institute's worldwide Innovative City of the Year award in 2013, beating New York and Tel Aviv in the process.

The spark for these improvements and the growth that has followed came twenty years earlier, however, when Escobar, listed in the 1980s as the seventh-richest man in the world, controlling 80 per cent of the global cocaine trade, was tracked down and killed by the authorities in a Medellín *barrio*.

Complementing security actions

Escobar was hunted by the authorities for six years, the noose tightening over the last six months of his life. 'His family was living in Bogotá, and he missed them', recalls one police intelligence officer. 'So he contacted them regularly on a satellite phone'.[6]

With US-supplied electronic surveillance narrowing his presence down to three Medellín *zonas*, the end of the man known as the 'King of Cocaine'—worth an estimated $30 billion at his peak—came shortly after a lengthy telephone call to his daughter. Escobar was shot ignominiously and alone on a slate-topped roof, fleeing his anonymous middle-class house. The once all-powerful 'grand capo', at one time an aspiring president, had just one bodyguard with him in his final battle.

Escobar's end, like that of fellow Medellín drug lord José Gonzalo Rodríguez Gacha, nicknamed 'El Mexicano' and killed four years earlier, signalled the advent of a new security and intelligence regime, and a renewed war on drugs. This was intensified by the increased funding link between FARC and the drug trade (discussed in Chapters 1 and 3) that emerged during the 1990s as the paramilitaries relinquished space and the guerrillas sought alternative means of income as Cold War sponsors evaporated.

It was the beginning of the end for the large Medellín cartels, which have since, under growing security pressure, 'atomised' into second- or third-tier leadership 'ODIN'[7] (Organizaciones Delincuenciales Integradas al Narcotráfico, Delinquent Organizations Integrated into Drug Trafficking) and still smaller 'combos', essentially joint ventures between druglords. By 2015, there were two major criminal structures in Medellín—Clán Úsuga and Los Rastrojos—controlling an estimated eight ODIN and more numerous combos.

The tables have now been turned. Whereas the 'commercial' life of a drug trafficker was estimated in the 1970s and 1980s at between fifteen and twenty years, according to police intelligence, by 2015 this had reduced to just twelve months. And whereas the cartels were focused on international markets and involved in the range of farmer-to-consumer activities including harvesting, production, transport and sales, today no single group can manage this entire range of activities.

Policing is now controlled from Medellín's high-tech dispatch centre located in the mayoral offices, where police officers monitor feeds from giant television screens. Such a whole-of-government response is but one aspect behind bringing down crime. The creation of 4,781 police *cuadrantes* (quadrants), or precincts, in 1,092 of 1,113 Colombian cities has helped to bring the police officer closer to the citizen. Officers make regular patrols and call in on each and every household, and the latter have the cell numbers of the patrolmen to hand. This system feeds back into better intelligence. Improved citizen-police co-operation is complemented too by police-sponsored initiatives such as the Cuadrante Amigo cell-phone app, which uses geospatial data to pinpoint users' position and, when activated, calls the nearest police patrol to their aid. There are always two officers patrolling each quadrant, on eight-hour shifts.

'These personal relationships', says General José Mendoza, police commander of Antioquia, 'are designed to prevent criminality. We no longer need a Rambo police officer, who was similar to the army officer'. Instead, he reflects, 'we need to prepare for peace, which means we need to deal with social and not just military conflicts'.[8] The police, too, recognise that it is very difficult for them to fight corruption and extortion if they themselves are seen to be carrying out such actions (even though allegations of such activity persist, especially outside the cities).

The capabilities of the police, too, have risen as their standards of training have improved, and as force densities have increased. In March 2015 there were 10,211 police officers for the estimated 3.5 million *paisas* (citizens) of the wider Meval metropolitan area around Medellín,[9] almost double the number there were in 2000.[10] Before this, as Jorge Giraldo-Ramírez and Andrés Preciado-Restrepo note, 'From 1990 to 1995, among the ten most important cities in Colombia, only Córdoba and Magdalena had less police per 10,000 inhabitants than Antioquia, and only Cúcuta and Barranquilla had less manpower than Medellín, which was well below the national average, notwithstanding the fact that the city continued to be the most violent in the country.'[11]

These force numbers must be considered along with the improvement in the quality of the police, notably in the percentage of graduates,[12] and the extent of cooperation with the military. Police activities have been closely integrated with those of the 4th (military police) battalion of the 11,000-strong 4th Brigade of the army, headquartered in Medellín.

While the city is best known for its drug cartel, it was also a focus of FARC and ELN activities by the late 1990s. Indeed, three initial objectives were set by then-President Andrés Pastrana as he sought to turn the faltering security situation around in 2000. These were to clear Bogotá of guerrillas; stabilise the eastern part of Antioquia around Medellín (where much of the country's electrical power is generated from hydro-electric plants and where as many as 100 pylons per night were being blown up by FARC); and disrupt guerrilla operations in the south-east around Macarena and Meta. While FARC and the ELN never made great inroads into a city once dominated by the paramilitaries and their self-defence groups, the strength of the drug trafficking organisations demanded a coordinated military-police response, consolidated through the mayor's office. Like the police in the quadrants, the 815 men of the 4th battalion of the 4th Brigade work in mobile units and in a range of activities including civic affairs, strengthening the judicial and health-care systems.

There are deeper challenges that go beyond homicide numbers. As noted by Colonel Herrera of Medellín's police:

> The culture of drug trafficking prevails in the form of a way of life. People have integrated themselves, and their families, into this chain, and see it as a legitimate way of life. It gives them status in the community and protection, and offers easy money. We thus have to deal not only with the symptoms, but also the causes, within the family ... [amongst] young men in particular, and with changing attitudes ... [to] work.[13]

Medellín's success is, in this regard, linked too with the demobilisation of the paramilitary blocs that formed the Autodefensas Unidas de Colombia (United Self-Defence Forces of Colombia, AUC), starting in 25 December 2003 when 874 AUC members reintegrated into society. There has been a proportional relationship between demobilisation and the decrease in homicides in the city, 'to the point that areas with a lower percentage of reintegrating people also had the lowest variation in the homicide rate'. Coupled with improved policing and governance, the results have been astonishing. Conversely, crime has risen where there have been disputes between drug trafficking groups, where there is corruption among judicial authorities,[14] and where there has been a breakdown in government relationships, such as occasionally between the city and Bogotá.[15]

A history of growth

Colombia's approach to economic policy and management is best described as one of 'gradualism'—of slow, incremental change. Despite a long tradition of political violence since the nineteenth century, 'no-one', former President César Gaviria (1990–94) observes, 'has ever questioned the viability of the country'.[16] In the last hundred years, the Colombian economy has contracted only during three years—1929, 1930 and 1999—and has delivered consistently high average rates of growth.[17]

This gradualism can be seen in the manner in which the government—and its citizens—have managed boom periods. 'Back in the late 1940s', remembers Andres Escobar, the deputy minister of finance under President Juan Manuel Santos, 'when coffee became important and the producers became the ruling class, a system was created isolating the country from the extremes of the mar-

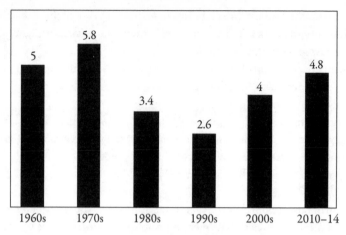

5. Colombia's Historic GDP Growth, %

ket through a savings fund which was given to the government to manage'.[18] While the rapid growth in the coffee industry had led to unprecedented economic prosperity, this had contrasted with widespread political turmoil. La Violencia—the ten-year civil war fought from 1948 to 1958—displaced whole communities and left hundreds of thousands murdered as competing systems of political patronage, more than ideology, split the country between Liberals and Conservatives. 'The economy is going well', became the standard refrain during much of this period, 'but the country is going badly'.[19]

Later, in the early 1990s, things changed, for two reasons. First, a new constitution established under President Gaviria massively expanded the welfare state, but without the tax base to support it. The fiscal deficit ballooned to 7.5 per cent of GDP and the economy collapsed in 1999. Then-President Pastrana was thus forced to go to the International Monetary Fund and implement a fiscal adjustment plan.

Second, the impact of drug trafficking on governance and crime began to bite. The expansion of the drug trade epitomised by Pablo Escobar was connected with the 'lost decade' of economic activity across Latin America in the 1980s, when growth came down from the 5 per cent average that had been sustained almost uninterruptedly since the Second World War. Colombia's annual GDP growth fell to an average of 3 per cent between 1980 and 2000. Mauricio Cardenas, formerly head of the Fedesarollo economic think-tank and minister of finance under Santos, says that 'the expansion of drug-related activities ... divided our history and changed Colombia'.[20]

The success of eradication programs in neighbouring Peru and Bolivia and the relative absence of state authority in the isolated savannah and jungle areas to the east and south of Colombia's Andean region—not incidentally, the remote periphery discussed in the last chapter, which formed FARC's heartland—set the stage for the cultivation of coca and a drug boom in Colombia. This also fed off the smuggling mindset and expertise that had developed out of protectionist Spanish colonial practices, and had been fomented during subsequent eras of oligopolistic control. It was no coincidence that Pablo Escobar started his criminal career managing a network stealing and smuggling gravestones to Panama and smuggling bicycle parts back, before switching to drugs. Not only did Colombia possess a comparative advantage in contraband operations, but the lines distinguishing law and order, good and bad, political, drug-related and social violence, had become blurred. From his support for Colombian cycling, to free public entrance to his Medellín Zoo and finance for local football, Pablo Escobar courted the patronage of the poor. His sponsorship of the 'Medellín without slums' campaign, an effort to build 800 bungalows for those subsisting on a refuse tip, made Escobar a cult figure even after his death.[21]

Indeed, at the beginning, there was a perception that cocaine-generated revenues would make all Colombians better off. Soon, however, given the impacts on governance and the slide towards criminality, tolerance was replaced by a realisation that drugs had corrupted Colombian society, undermined the economy and judiciary, and had a negative impact on productivity along with human and physical capital—key components of economic prosperity.

Against this backdrop of failing security and a weakening economy, however, in 2002, newly-elected President Álvaro Uribe offered a break from the past, not only by breaking the alternating duopoly of Liberal–Conservative domination of politics as an independent, but also by taking a grip of the security problem in a manner unlike any of his predecessors. In so doing, he was able to create a wave of national political support and foster an unprecedented degree of consensus in the country in support of change and progress. 'He flipped the Colombian psyche',[22] said one diplomat based in Bogotá. 'Until Uribe, the lesson learned was not to stick your neck out and take risks. But he was a man for whom there was no dimmer, just an on-off switch, and it was mostly on.' He changed matters by his energy, attention to detail, and by holding the military and his ministers effectively to account.

Whereas his predecessor Pastrana was elected to try and find peace, Uribe had, to his peril, been elected to take the fight to the guerrillas and paramilitaries. Uribe also emphasised the relationship between economics and security:

> Colombia performed very well during the Latin American economic crisis. Colombia never defaulted. Colombia never had hyper-inflation ... Colombia had always fulfilled all the financial obligations of the international community. However our investment rate was very low. Unemployment was between 16 and 21 per cent. Poverty ranged between 47 and 53 per cent. Yet Colombia is a country with a great entrepreneurial spirit and a great past in economic performance. But we needed security ... and confidence.

Uribe highlighted this in reference to his promotion of 'three things to restore confidence: security with democratic values, investment promotion and social cohesion'.[23] His policy enshrined security and the strengthening of government institutions as the means to improve governance and bolster development; without security, he stressed, the other objectives of government could not be accomplished.[24]

As Uribe's vice-president, former journalist Francisco 'Pacho' Santos put it in 2006, 'security and stability is a crucial element of our growth. As the government has started to regain the monopoly on the control of violence, the economy has boomed. This is not the result of a boom, as elsewhere, of oil or other commodities, but a boom in security.'[25]

Pacho Santos claims that Uribe's successor (and his own cousin) Juan Manuel Santos had 'given it all up',[26] not least by 'starting the peace process which put the FARC at the same level as the army', allowing the guerrillas to 'gain ... space not through hard-core terrorism but through extortion and other means', and by losing the trust of investors. Other supporters of Uribe say that politics is now, as a result, driven at the local level by clientilism, not least because 'the current government arrived with money rather than reforms, focusing its policies on interest groups and subsidies, losing the fiscal space in the process'.[27]

In fact, aside from issues of tone, there is little to choose in economic policy terms between the Uribe and Santos administrations. Both have pursued the same commitment to fiscal balance, sound monetary policies, and the promotion of external trade and investment. If anything, the Santos government has gone further by reforming the rules on the distribution of oil and gas royalties, ensuring these are spread more widely, in the process constricting a source of support for the guerrillas from among the oil-producing areas.

The Santos administration has instituted other novel reforms to pump up growth. In 2013, the government delivered low mortgage rates through a public subsidy and an agreement with banks—$200 million of public money was used to leverage housing investment of $2 billion, boosting construction. A 2012 law cut payroll taxes while raising income tax on the better off, growing formal-sector jobs at 8 per cent per year, and boosting productivity in the process.[28]

Importantly, under both Uribe and Santos, the improvement in security has enabled the economy to develop; what Cardenas has described as a 'reversal of fortune'.[29] Sound macro-economic policies have supported the sustained growth of the economy, including the adoption of an inflation-targeting monetary commitment in 1999, a prudent spending regime as articulated in the 2011 Fiscal Rule Law, as well as market-friendly and consistent policies.

Along with a reduction in public debt levels (to 1.5 per cent of GDP in 2014), a more than doubling of credit availability to the private sector (from 17 to 30 per cent of GDP between 2004 and 2014) and an export-led growth strategy, Colombia's growth rate has steadily risen while poverty has fallen by 20 per cent. Unemployment is down markedly from 15.7 per cent in 2002 to 9.1 per cent in 2014.

In the process, Colombia has jumped one place to become the fourth-largest economy in Latin America, behind Brazil, Mexico and Argentina, but now well ahead of troubled Venezuela. The government's economic policy and Democratic Security Policy have engendered a growing sense of confidence in the economy, particularly within the business sector—a confidence encouraged by fiscal incentives (essentially tax holidays) worth as much as 1 per cent of the country's GDP of some $382 billion by 2014.

Change has not only been about financial figures and riches, however. In 1993, national health coverage reached just 30 per cent of the population; by 2015 this had been extended to 'essentially 100 per cent', according to Andres Escobar.[30] Education coverage went from 70 per cent to 94 per cent over this time.

Additionally, a wider social programme includes cash grants to the poorest citizens and focuses on housing, among other needs. The conditional cash transfer system, Familias en Acción, was established in 2000. These transfers provide monthly grants of up to $28 to poor households with children, provided they meet certain conditions, including the requirement that children aged under seven attend regular medical check-ups, and that children aged between seven and eighteen attend no less than 80 per cent of school classes during the school year. By 2006, transfers reached 400,000 households (or 5 per cent of the population) in 700 municipalities, with an annual budget in 2004 of $95 million.[31] Administered by the Departamento para la Prosperidad

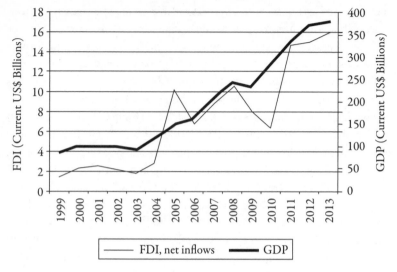

Source: World Development Indicators.
6. Colombia: FDI and GDP

Social (Department for Social Prosperity, DPS), by 2015 the programme was present in 1,102 municipalities, reaching 2.6 million families, and with an annual overall budget of $800 million.[32]

Economic growth has been driven, on the one hand, by increased public spending on infrastructure, and on the other by an increase in commercial investment. The ratio of investment to GDP has nearly doubled, from 16 per cent in 2000 to 29.8 per cent in 2014, reflecting better business conditions and, again, improved security—both aspects reflected in the regaining of investment grade ratings lost in 1999. Portfolio investments also increased from an average monthly withdrawal of $1 million between 1999 and 2002 to an average inflow of $1 billion by 2015, while foreign direct investment (FDI) has doubled to 5.1 per cent of GDP between 2000 and 2015. Much of this has been on the back of the commodity super-cycle and high oil prices of the early twenty-first century.

Regardless, a positive economic–security cycle has been established and has helped. Uribe's former finance (and, earlier, agriculture) minister, Dr Roberto Junguito Bonnet, has estimated the components of the economic growth witnessed: 'I believe that around one-third of the increase in economic growth has been because of improving security, one-third because of the good international environment, and one-third due to good economic policies.'[33]

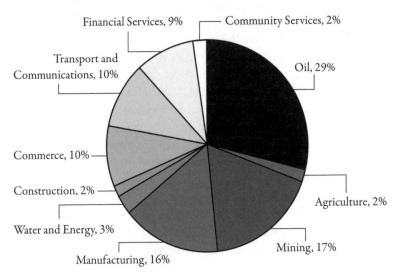

Source: Colombia Reports.

7. Colombia: Inward FDI by Sector, 2013

Around half the increase in investment has originated from oil companies, on the back of key reforms in 2004 aimed at deregulating the sector. But there have been others.

For example, SAB Miller, the world's second-largest brewer, made a $7.8 billion investment in 2005, to date the largest in the country, in purchasing Bavaria Breweries from the Santa Domingo family. Aside from the commercial imperative to acquire the world's tenth-largest beer producer (and one with an astonishing 99 per cent market share in Colombia, Ecuador and Peru) with very profitable operating margins, the wider drivers for this have been acknowledged as the positive and dramatically improved security situation, Colombia's strong democratic traditions, the presence of free-trade areas, including with the US and the European Union, and, overall, a business-friendly, politically stable environment where the rule of law is respected and where growth has remained strong despite the prevailing international headwinds. Additionally, a number of incentives were available to SAB Miller, not least a tax stability agreement. As operating efficiencies have improved, Bavaria has become by far the most profitable business in the SAB Miller stable. And the government remains a receptive and willing partner—for example, it has assisted in the development of a barley sector within Colombia to supply the brewer.

Also beyond oil, AngloGold Ashanti's operations in Colombia hint at the efforts made by government to facilitate other forms of investment, with exploration by the world's third-largest gold company starting in 2002. As a measure of this industry's long-term horizons, by 2015, AngloGold Ashanti had invested more than a quarter of a billion dollars in various green-field projects in the departments of Tolima and Antioquia. Aside from being in a dynamic growth environment, in the words of AngloGold Ashanti's local vice president, a combination of 'good policy and a co-operative mining agency and the way in which government expedites investment'[34] has made Colombia an attractive destination for such long-term investment. All of AngloGold Ashanti's projects have been among the eighty-four designated by the government as Projects of National Interest (PINES), enabling a single-window approval system, 'meaning that you don't have to deal with sixteen different entities'.[35] Still, there remain challenges in linking the concerns and constituents of local municipalities with the intent of national government. A related need is to manage the extent of artisanal, and often illegal, mining, especially as the country transitions to peace.[36] The issue of linking the centre with the periphery—discussed in detail in the previous chapter—has been an age-old feature of Colombia in economic as much as in political or security terms.

Exclusion, inclusion and the 'two Colombias'

It was competition among Colombia's elites that led, in 1946, to La Violencia, a violent struggle between Liberals and Conservatives costing around 300,000 lives. The power-sharing National Front pact that ended the conflict, however—establishing the alternation of government every four years between Liberals and Conservatives—simultaneously disallowed the participation of other political parties. This sowed the seeds for the guerrilla conflict that followed as divides widened between wealthy landowners on the one hand and rural *mestizo* (mixed race) and indigenous peasants, two-thirds of whom lived in poverty in the 1960s, on the other. As a result, FARC's original aim, upon its creation in 1964, was to overthrow the state and redistribute land and wealth among the entire country at the expense of the elites.

This is what is we have described as the problem of the 'two Colombias'. This phrase speaks to the paradox of Colombia, with its top-ten status as a tourist and investment destination, and its poor rural areas and inner-city deprivation—the difference, for example, between the financial districts of the cities and the unregulated situation of many rural towns. There are a number of measures of such informalisation: for example, less than half of the Colombian population of 48 million is banked, and the tax-to-GDP ratio is under 14 per cent, compared to the Latin American average of 19.4 per cent, or 33.8 per cent among Organisation for Economic Co-operation and Development (OECD) nations.[37] This hints at a weakness in governance extension, a lack of faith in government institutions, the presence of large informal and illicit economies, and widespread poverty. These problems run deep.

As Harvard University's James Robinson notes:

> It is tempting, and common, to attribute such problems to the drug industry, but this is a mistake. The country's status as the capital of world drug trafficking reflected the prior dysfunctional organisation of Colombian society. It is also tempting, and wrong, to blame Colombia's woes on the guerrillas. Like the drug industry, they are an outcome of more deep-seated problems.[38]

Fundamentally, he argues: [A]ll the ills that Colombia has experienced stem from the way it has been governed. The best way to conceive of this is as a form of indirect rule, common during the period of European colonial empires, in which the national political elites residing in urban areas, particularly Bogotá, have effectively delegated the running of the countryside and other peripheral areas to local elites. The provincial elites are given freedom to run things as they like, and even represent themselves in the legislature, in exchange for political support and not challenging the centre. It is this form of rule in the periphery that created the chaos and illegality that have bedevilled Colombia. Drugs, mafias, kidnappers, leftist guerrilla groups, and 'rightist' paramilitaries certainly have exacerbated the country's problems, but the problems all have their source in the nation's style of governance.[39]

This system has been perpetuated historically, Robinson contends, by the fact that the turmoil in the countryside 'lowers the price of votes', meaning that politicians get elected by winning the support of local bosses rather than fighting elections on the basis of platforms and promises to the electorate—'or perhaps [by] becom[ing] ... the bosses themselves'.[40] The elites, he argues, have cooperated to guarantee a share of power without costly investments in rural areas, local opposition has been divided and ruled, often through extreme violence, while the concerns of these areas were maintained far to the periphery of Bogotá's vision. This, along with the fundamental mismatch, in population terms, between Colombia's centre and its periphery, is the economic and political underpinning for the territorial logic of the guerrilla stalemate we noted in Chapter 3.

Unsurprisingly, politicians and others seek resonance in this environment of exclusion and inclusion, outsiders and insiders, a two-speed 'mixed-opportunity' economy. While FARC maintains a broader political agenda—or its senior leaders speak as if they do—in practice, the behaviour of FARC fronts has degenerated into drug trafficking and extortion, among other forms of brigandage, and thus land reform is one of the few political issues they have continued to flag.

There is some justification behind this. Land distribution in Colombia's rural periphery is highly unequal, with 52 per cent of farms in the hands of just 1.15 per cent of landowners, according to the United Nations Development Programme. The levels of inequality are linked to issues of distribution. Only one in two peasants possess title to their land, and rural property is highly unequally distributed.[41]

These divides resonate in high levels of income disparity within the economy, reflecting patterns of access to education and infrastructure, among other aspects. And they perpetuate themselves. Colombia remains one of the most unequal societies worldwide as measured by its Gini coefficient, though there has been a steady decline in this Gini number, as measured by the World Bank, from 55.5 (2000–04) to 54.2 (2005–09) and 53.5 (2010–14).[42]

Such problems were recognised by the Uribe government in its National Consolidation Plan (NCP) in 2007, and subsequently, too, by the Santos administration, encapsulated in the ambitious Victims' Law signed into effect in June 2011. This law aimed at resolving past conflicts by restoring vast

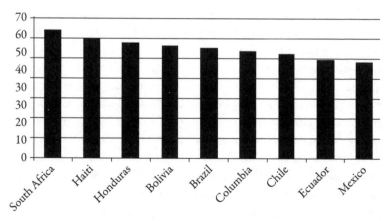

8. Inequality: Gini Coefficient, 2012

amounts of land to people who had been dispossessed in the fighting, affecting as many as 4 million people and 5 million hectares. Still, the rollout has not matched the rhetoric. The government has been constrained by problems of security in the rural areas—demanding the kinds of militarised development responses discussed in previous chapters—and by a lack of money. Funding for the NCP, meanwhile, has come from the US Agency for International Development (USAID), which had committed $237 million by 2015.

In part, these issues are also about attitude. The citizens of Colombia's cities have traditionally been unconcerned by the guerrilla threat. Not until it threatened their own security in the late 1990s, when FARC developed semi-conventional capabilities and approached Bogotá, did they elect a government that pledged to do something about it.

Bridging this gap in wealth and opportunity thus requires, in Colombia as in other developing economies, on the one hand efforts to deal with uncompetitive monopolistic and oligopolistic behaviour; and, on the other, the creation of conditions in which employment opportunities, among other possibilities for wealth creation, are widespread. The elites have traditionally monopolised different sectors, from banking to brewing, sugar and services. Such control served to distort domestic pricing, inevitably to the disadvantage of Colombia's poorest. One effect has been a flourishing trade in contraband.

The mayors of Medellín

Pablo Escobar's stomping ground, Medellín, offers one illustration of how progress might be made in the face of considerable adversity. While the rate of poverty has been reduced substantially to less than 40 per cent, and unemployment has fallen officially to around 10 per cent, the city remains the most unequal in Colombia, with a Gini coefficient of 0.54. As Aníbal Gaviria, the mayor of Medellín, has put it, 'Inequality is the biggest challenge facing Colombia today'.[43]

In attempting to address the city's deeply rooted problems of poverty, inequality, informality, social injustice and violence, one response has been to spread credit via micro-finance through a network of business-development agencies known as Centros de Desarrollo Empresarial Zonal (Zonal Business Development Centres, CEDEZOs).[44]

Sergio Fajardo was the mayor of Medellín from 2003 to 2007 and has been the governor of Antioquia since 2012. As mayor, he set out to solve three problems, says his Government Secretary Santiago Londoño: 'violence, inequality and a culture of illegality'.[45] To achieve these aims he attempted to 'build a bridge between the poor and rich neighbourhoods' by providing parks, libraries, schools and security in poorer areas, offering 'the best for the poorest people', alongside the CEDEZOs system of micro-loans. These policies were informed by the programme of Securidad Integral (integrated security)—in Londoño's words, 'a mixture of politics, human rights, reconstruction, rule of law, reparations, institutionality and crime prevention'. He cites as an example of its success the progress made in Medellín's Comunas 1 and 2 between 2004 and 2008, with the number of businesses surging from just thirty-eight to 250 in only four years.

The funding for this programme—some $2 billion annually in Medellín alone—has come from the Empresas Publicas de Medellín (EPM), a 100 per cent publicly owned utility which manages energy, water and gas resources.

Graduating these small businesses (known as *chasitas* after the small trolleys the poor use to move around the community to hawk items) into something more sustainable and prosperous has, however, proven more challenging. Regardless, Fajardo has taken this approach into his governorship, extending it across 125 municipalities into rural areas where, as Londoño explains, 'the FARC and ELN, unlike in the city, are very much still alive'. 'Most of the region,' he says, 'has been stateless, where the only state institution the citizens encounter is the army, which most find difficult to relate to in its counterinsurgency role'. In these circumstances, 'it is impossible to end conflict and crime unless one can offer a different lifeline to the youth ... Instead of joining BACRIM or the guerrillas, they need to get a job'.

Fajardo has just 60 per cent of the city's budget to play with across his 63,000 squared kilometres department. No longer able to call the shots directly, he has to work through the mayors, and around the very weak regional infrastructure, especially the road network. Not all problems centre on objective conditions. 'We have encountered a lot of resistance from the people in the city who question why we would give the guerrillas concessions and incentives—they simply don't understand it,' reflects Londoño.

A wider issue

Medellín's challenges in moving from subsistence to jobs can also be seen in the central government's sometimes-fraught attempts to diversify the economy. As a result of the pursuit of an import-substitution model Colombia, like most of Latin America, developed a range of industries from the 1950s until the early

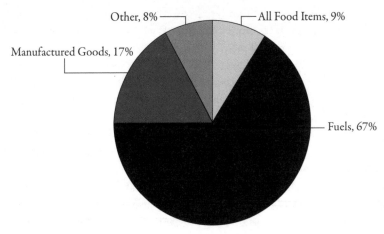

Source: UNCTAD.

9. Colombia Exports 2013

1990s. Yet as the trade environment opened up in the early 1990s, and faced with the challenges of a lack of domestic scale and the costs of weak infrastructure, many of these industries have failed to prosper or survive.

As a result, natural resources and commodities—including coal, oil, coffee and cut flowers—together comprise around 70 per cent of Colombia's exports. The government has swung 'like a pendulum', says Andres Escobar,[46] in trying to achieve competitiveness, between intervening directly on the one hand and, on the other, trying to provide general public goods and a conducive environment for business. For example, in 2007 the Colombian government launched a 'Productive Transformation Program' to create public-private partnerships across several industrial sectors in seeking higher-value addition: business process outsourcing and off-shoring, software, cosmetics, personal-care products, health tourism, textiles and clothing, electricity, auto parts, and printing and graphic arts.[47] The results were mixed, however, and by 2015 the government had moved back towards a more liberal, hands-off model, focusing less on picking sectoral winners than on creating an enabling environment though its '4G' programme to build roads countrywide, the benchmark for success being for trucks to travel at 80 kilometres per hour on every major road. Estimated to cost $20–25 billion over ten years, the scheme was to be funded by public–private partnerships.

* * *

There is no simple answer to meeting the challenge of the 'two Colombias'. This is just one among the many challenges of moving from war to peace, and from a highly unequal and racially stratified society to one defined by equal opportunity. In the past, the scarcity of avenues for social mobility provided a constituency for radical insurgents and populist sentiment—or, at least, the motivation for those making a living from illegal means.

In part this has to do with the way in which the country is governed. Colombia's historical dysfunction exists in the distribution of governance and finance. Municipalities are ranked, for example, from one to six in terms of their population size and income, where Bogotá is a 'one' and poor, rural municipalities are a 'six'. The higher the rank, the greater the income, and the greater the prospects for governance, fuelling a vicious cycle of underdevelopment. But in the rural areas, where the property market is under-regulated and tax income is low, finances for a proper bureaucracy and infrastructure are limited. Foreign diplomats estimate that as many as half of all Colombia's municipalities fall into the sixth-level category.[48]

And partly this is down to opportunity, skills and expectations. While offering a promising alternative—a means of decent income—to those left out and behind is the long-term solution to insecurity, this is especially difficult when, after fifty years, generations of people—known generically as *familias farianas* (FARC families)—have become deeply inculcated into a guerrilla or criminal way of life and means of survival.

Far on the outside

As discussed in Chapter 1, FARC originated in the Sumapaz area of southern Tolima, 90 kilometres south of Bogotá. In this area, 3,000 metres up in the eastern Andes, FARC massed 2,000 soldiers in 1999 in anticipation of its planned onslaught on the capital. Mono Jojoy himself grew up in the area, in the small town of Cabrera.

By 2014 the 3,431 men of Joint Task Force (JTF) Sumapaz were in control, and FARC is down to an estimated 200 guerrillas in an area of 370,000 people stretching across twenty municipalities and 6,600 square kilometres. FARC has hung on, says the JTF commander Colonel Erik Rodriguez, because their support networks (discussed in detail in Chapter 3) are deeply ingrained in the area and among its people, and because they need to maintain a presence—a force-in-being—to ensure political clout at the negotiating table.

In March 2015, the JTF was busy building a sports centre at Cabrera under its Accion Integral programme, a $100 million three-year fund to build essential infrastructure across 142 of the most conflict-afflicted municipalities countrywide, the aim being to enhance security, isolate communities from the guerrillas, create the perception of a government presence, and improve quality of life. In the case of Cabrera, it is a way of saying, too, that the era of Mono Jojoy and his type is over and that the government is here to stay.

In the centre of the town of La Macarena, 280 kilometres to Bogotá's southeast, a mural painted by FARC behind the altar of the local church depicts a yellow-shirted guerrilla holding a priest hostage, seated next to a saintly Jesus. The guerrilla's gun has been airbrushed out, but the mural is retained as a reminder of the FARC era.

Today, horses range freely in the town square, mowing the grass in front of the church, and on the main road is a collection of small restaurants, shops, and vendors selling fresh juice and sugar cane, Colombian football shirts and other basic goods. At the end of the road is the Guaviare River, its fast-running brown waters swirling through the town from San José del Guaviare through FARC-infested territory. Along its banks are small shacks, the marines regularly patrolling its reaches in small, heavily armed boats.

With 6,000 people in the town and another 25,000 spread over 196 villages across the area, income in La Macarena comes mainly from 5,000 ranches totalling 120,000 head of cattle, and from selling services to the 7,000 troops based in the area. The nearby Caño Cristales (Crystal Canyon) in the Sierra de la Macarena National Park, described as the 'eighth wonder of the world', where clear water flows over plants creating shades of black, green, white, yellow and ruby red, is becoming an international attraction. From just 700 tourists in 2009, this sideline grew to 6,000 visitors by 2014, including that year Britain's Prince Charles, leading to a mini-boom in guiding, hotel accommodation and restaurants.

'Thanks to God and the military forces', says the town secretary, meeting under the shade of a thatch in June 2014 at Joint Task Force Omega in La Macarena,[49] once the 'capital' of FARC's Caguán demilitarised zone, 'we have been able to recover the area. The greatest desire of Colombians is to be free of the FARC scourge.' However, as his colleague, Councillor Hugo Morena, notes, 'Recovery takes more than money'. The most important statistical indicator, Morena and his colleagues concur, is education. 'When this was a demilitarised zone, there were lots of bars, and armed and drunk kids. They had money, but that's not development' said one.

Still, the scale of financial resources available for La Macarena's development—and their military source—is indicative of the challenge of maintaining stability without troops. In June 2014, there was $750,000 in the municipal budget, of which just 15 per cent was derived from local taxes. All capital projects require investment from the military. La Macarena's mayor acknowledges that he is a 'man looking for money from central government'. Graduation from the protection of the state for security and income will occur when, as the then defence minister Juan Carlos Pinzón put it, 'a person has a regular income and a job. The challenge is that the only jobs in many of these areas are in the military, police or in the mayor's office.'

This applies to other areas of Colombia's periphery—which, as we have seen, is the frontline of the fight against both FARC and poverty.[50] One of the poorest of Colombia's thirty-two departments, Norte de Santander, nestling in the north-east on the border with Venezuela, has historically played host to virtually everything among its 1.2 million people apart from good government: left-wing guerrillas, right-wing paramilitaries, coca growing, illegal gold and Coltan mining and narco-trafficking. Drugs and insurgency have become inextricably integrated with the region and with each other during the past fifty years of trouble. The years 2011–12, for example, saw more than 400 drug laboratories and nearly 80 tonnes of cocaine destroyed in Norte de Santander alone.

This region is sparsely populated and lacks infrastructure. The jungles, frequently bisected by curving, wide, mud-brown rivers, occasionally make way for parcels of oil palms, though farming is defined more by occasional cattle ranches, smallholdings and secret coca plantations. It is, said Pinzón during a visit there in November 2013, 'the heart of Timochenko country',[51] referring to Rodrigo Londoño Echeverri, alias 'Timochenko', who took over as head of the FARC Secretariat on the death of Alfonso Cano in 2011.

The bridge over the sweeping Catatumbo River would not be out of place on the set of David Lean's 1957 classic *The Bridge on the River Kwai*. The structure's age and poor design tells its own story of the government's paucity of resources and an earlier age of ignorance and neglect. 'These are people that the country forgot', reflected the then Vice Defence Minister Jorge Enrique Bedoya.[52] The absence of roads, the extreme topography and high levels of insecurity, for example, meant that the town of La Gabarra was accessible only by flying in a UH-60 Black Hawk helicopter from the airfield at Tibú 60 kilometres away.

But these citizens are no longer being ignored. La Gabarra and the surrounding region has become the epicentre of the government's strategy to win control of the countryside, deal with the narco-trafficking and insurgent threat and, in so doing, extend governance to all.

This has demanded keeping the pressure on FARC. But the strategy has had its limits, especially among populations caught between the violence and promises of the guerrillas and the reality of a lack of economic opportunities. Poverty is widespread, and opportunities few and far between. In La Gabarra, a one-horse town if ever there was one, the same scruffy, cheap restaurants and 'fashion' shops line the main dirt road, one after another. Pinzón was in La Gabarra in November 2013 to open one of eighteen new sports centres built countrywide by the army in 2013—this one by the 30th Engineer Battalion, 'Jose A Salazar'. In dripping humidity and 35-degree Celsius heat, exacerbated by the tin-roofed complex, the whole town seemed to turn out for the occasion, with even the local talent-for-hire lining up and leering from their balcony as the soldiers patrolled by. Pinzón's message was clear. The government, not FARC, has built the sport facilities and is building the road to Tibú. If any of you, he told the crowd, by then spilling out of the complex, knows someone in FARC, a friend or a relative, let them know that they can demobilise.

When the mayor got up to speak, the questions from the crowd in La Gabarra spoke volumes about the intersection between governance, security and economic opportunity. Residents asked for a bank to be established (there was not one in the town) so they could pay their debts. Another wanted a garbage service for the town. The then Chief of the Armed Forces General Leonardo Barrero empathised with their plight, referring to his past experience in managing coca eradication elsewhere, encouraging the farmers to get into other crops and become part of the solution. All of this, however, costs money.

Funding the peace

Economic growth has expanded the slice of funding available to the armed forces to prosecute the war on FARC and on criminality. The government has also successfully employed financial instruments as tactical instruments in funding the war on the insurgents and, perhaps even more importantly, in building a strategic consensus around the increase in security expenditure. The defence budget has increased from $3.1 billion in 2000 to $13 billion in 2013, though it has remained consistent, given the growing economy, at around 3.5 per cent of GDP since 2002.

These figures require disaggregation. The defence budget includes spending on the 180,000-strong police force, along with the 285,000 men and women in the armed forces, given that the National Police fall under the Ministry of Defence's purview. Of the 3.3 per cent slice of GDP received in 2015, for exam-

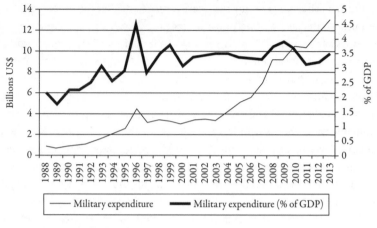

Source: MINDEFENSA.

10. Colombia's Defence Spending

ple, 0.7 per cent went to pensions, 0.5 per cent to the administrative 'back-office', 0.9 per cent to the National Police, and 1.2 per cent to the military.[53]

The spending boost has been underpinned by improved international relations. Foreign support for the new security approach was procured under Plan Colombia, a $10.6-billion initiative to strengthen public institutions, especially the military. Presidents Clinton and Pastrana originally devised the plan in 2000 to finance efforts to curb coca cultivation by 50 per cent in five years. It had its origins well before that time, however, in 1962's Plan Lazo, a US-designed attempt 'to bring a social, integrated solution to the guerrilla problem',[54] which however foundered on the government's fear of a wider role for the military in an age of coups.[55]

Originally, the money procured under Plan Colombia was intended to be split 50:50 between the security forces and alternative livelihoods for rural *campesinos* (peasants), offering them a way out of coca farming. In practice, most of the money ended up being spent on the military and (at least initially) on American contractors for training, equipment maintenance and drug spraying. Appropriations under Plan Colombia made the country the third-largest recipient of US aid in the early 2000s after Israel and Egypt.[56] More important than the military aid, perhaps, was the shared intelligence available from improved relations after the upset of the era of President Ernesto Samper Pizano (1994–98)—whose US visa had been revoked on account of the controversy of campaign funding links between the president and drug-traffickers—as well as the knowledge that the US was again 'there' for the Colombians, or, as Pinzón has put it, that the two countries were again together 'on the bus'.[57]

Plan Colombia formally ended in 2013. The US financial commitment has continued, however, albeit at a lower level. In President Obama's 2015 budget proposal, Colombia was slated to receive $280 million in military assistance, a $40 million drop from 2014. Aid directed towards the fight against drugs in Colombia would fall to $117 million from 2014's budget of $142 million. US development aid was to drop to $133 million in 2015 from the $140 million allocated the previous year,[58] though the activities of USAID, as the largest foreign donor agency, were expected to continue to focus on four programme areas, namely land and livelihoods; demining, human rights, labour and electoral support; environment, climate and energy; and vulnerable persons, notably the indigenous and Afro-Colombian communities, as well as the reintegration of child and demobilised soldiers.

While operational, Plan Colombia was important both financially and politically to Bogotá. But it was not the only source of income. The increase in defence expenditure was partly financed through a 'wealth tax' on capital investments established in 2002, averaging out at around 0.6 per cent of GDP annually.[59] This source of funding served also to give confidence both to the US that the security surge was owned principally by the Colombians themselves and, according to Pinzón, to the armed forces, firm in the knowledge that the nation was behind them.

Local ownership has indeed been critical for success, but critical too in convincing Washington of Colombia's commitment. Certainly, the extensive US engagement has gone against the grain of political sentiment in much of Latin America at a time, from the 2000s, when much of South America lurched politically to the left and anti-American rhetoric abounded, most notably from Venezuela and its late president, Hugo Chávez. Against this backdrop, when in office, President George W. Bush described his Colombian counterpart, Uribe, as 'a strong and principled leader'. Bush said further: 'I admire his determination; I appreciate his vision for a peaceful and prosperous Colombia'.[60]

In November 2006, the two countries signed a bilateral free-trade deal, the biggest agreed by the US in the western hemisphere since the 1994 North American Free Trade Agreement with Mexico and Canada. This deal made 80 per cent of US exports to Colombia duty-free immediately, with the remainder to be phased out within ten years of the agreement coming into force in

2011. In 2014, the US was Colombia's biggest trade partner, with Colombia's exports worth $18.7 billion (nearly 32 per cent of Colombia's total exports) in 2013, and US imports reaching $16.4 billion (nearly 28 per cent of Colombia's total imports).[61]

Aid agencies, as noted above with regard to the National Consolidation Plan, have attempted to plug the government gap to meet the need for development expenditure in conflict-ridden areas. But there is little that they can or should do when faced with the lack of political will that has characterised these schemes—all the more remarkable since the overall success of the Colombian project since 2000 has been the result of such will, a laser-like focus on security and governance, technical capacity and budgetary wherewithal.

Started in La Macarena in 2005 and expanded to nine regions countrywide through the Centro de Coordinación de Acción Integral (Centre for Co-ordinated Integral Action, CCAI), the NCP quickly faltered due to a combination of insufficient funding, a lack of political heft once Uribe left office, and the re-imposition of Colombia's classic dynamic of stalemate between the guerrillas, BACRIM and the government on the ground.

By 2015, the NCP was operational in ten regions, involving fifty-eight municipalities deemed *zonas rojos* (red zones) by the Unidad Administrativa para la Consolidación Territorial (Administrative Unit for Territorial Consolidation, UACT), nominally under the government's Departamento

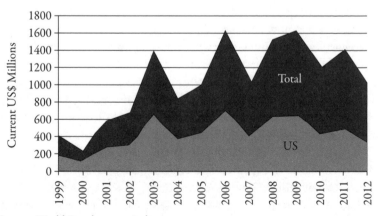

Source: World Development Indicators.

11. Colombia: Net Bilateral Aid Flows

para la Prosperidad Social, which is also tasked with working with the military's development programme, Accion Integral, in 142 other municipalities and in expediting the Familias en Acción cash-grant system. Although the government has claimed[62] that both these efforts and those in another ninety-six municipalities against illicit crops were 'complementary', foreign diplomatic observers assess that this has served to dissipate government effort, and at some cost.[63] Surveys show that between 2012 and 2014, for example, conditions in just 5 per cent of 3,500 *veredas* (hamlets) improved, while 4 per cent deteriorated. While civilian input is necessary to make full use of the clinics, schools and other facilities constructed, the military's evident frustration with the slow pace of civilian delivery led it to focus its engineering efforts instead on bridges, roads and sports-centres, which it could transfer to local authorities without the need for civilian staffing.

This failure has contrasted notably with the whole-of-government approach advocated during both the Uribe and Santos presidencies. As discussed in Chapter 3, the military's progress in counter-guerrilla warfare has not been matched by the sorts of civil agency programmes that are essential if wider counterinsurgency efforts are to succeed in targeting all aspects of the insurgency, or preventing the re-emergence of future conflict.

Overall, apart from freeing up fiscal space, the role of foreign aid as a tool for development is limited in Colombia, as in other developing and post-conflict environments, and its utility is overwhelmingly dependent on local leadership, ownership and capacity. The principal challenges to the Colombian political economy therefore centre less on providing micro-level spending than on managing the macro-environment, including the dependency on commodity incomes and the impact of this on the exchange rate and the competitiveness of other sectors.

The fragility link between security, growth and the global economy

Colombia has learnt that just as security is critical for growth, growth begets security.

The rise in global oil prices and concomitant increase in Colombia's oil production helped the country fund its security project since 2000. Since 2005, oil production has almost doubled from 525,000 barrels per day to around 1 million bpd. By March 2015 this had reduced slightly to 965 million bpd.[64] The oil sector attracted $4.3 billion in foreign direct investment, for example, in 2011, up from only $278 million in 2003, driven by higher global

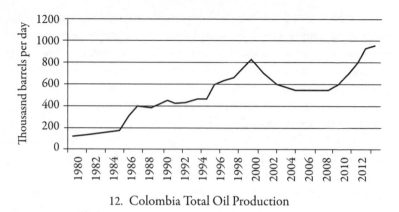

12. Colombia Total Oil Production

prices, tax breaks to investors, the opening of large areas for exploration, and partial privatisation of the state-owned oil company, Ecopetrol, also allowing international oil companies to wholly own upstream ventures.[65]

Conversely, the drop in oil prices from 2014 has posed challenges to Colombia's growth prospects according to the International Monetary Fund, causing growth to slow to 3.4 per cent in 2015 from 4.6 per cent in 2014. For every dollar that international oil prices reduce, Colombia forgoes an estimated $430,000 million pesos per year (or US$143 million, at the prevailing exchange rate in August 2015), according to a United Nations study.[66] As a result, Colombia's peso has been one of the poorest performing currencies during 2014/15, losing more than a third of its value.[67]

To an extent, however, Colombia has been shielded from the global commodity price decline by its strong policy framework—including its fiscal conservatism, inflation-targeting regime, flexible exchange rate, and a sound and stable financial sector. Regardless, the fall in oil prices has stressed the urgency of mobilising non-commodity revenues, at the heart of which is the need to broaden the tax base and facilitate private investment through improved competitiveness. Indeed, the fall in the peso's value should assist other export sectors, including horticulture and agriculture, and reduce imports into key sectors, such as automobiles, where imports account for two-thirds of sales. Without more inclusive growth, however, as in other fragile developing economies, issues of inequality, youth unemployment, informality of economic participation, weak social cohesion and insecurity will remain linked and a threat to national stability.[68]

The reforms thus required to ensure long-term competitiveness depend primarily on tough domestic political choices rather than on international largesse, as much as the latter might ease some of the extreme pain.

Necessary reforms centre on reducing income inequality, driven by unemployment and informality, which depends on international openness to trade and improved competitiveness—for example, through labour market reforms. Among its suggested reforms for Colombia, for example, the OECD maintains that the minimum wage should be differentiated by region, while also suggesting that the high level of social-security contributions and *parafiscales* (payroll taxes), which work against formal job creation, should be reduced, and the tax system made more progressive through removal of exemptions that mostly benefit the richest taxpayers. As Robinson reminds us, 'While the poorest 10 per cent of Colombians pay 8 per cent of their income in taxes, the richest 10 per cent pay just 3 per cent.'[69]

Finally, the need to boost productivity depends, as ever, less on outsiders armed with aid than on government policies that can promote productivity, notably through the education and training system, through greater investment in public infrastructure (especially in lowering the costs of getting to and from market), through reducing barriers to entrepreneurship (including improved access to finance), and through strengthening contract enforcement and eliminating corruption.[70]

Victory and graduation?

As the foregoing discussion makes clear, Colombia has done much right over the last fifteen years, in the security as much as the political and economic domains. The economy has benefited from steady growth for much of its recent history, with the uptick after 2002 both reflecting and enabling a security dividend.

This impressive record illustrates the need for determined leadership and attention to detail in following through in practice on visions and policies, and the importance of ensuring the macro-economic fundamentals, including low inflation and fiscal conservatism, spending on infrastructure and education, and attracting investment. Social peace can also be bought, to an extent, at least temporarily, through redistribution in the form of cash grants and infrastructure spending, through the NCP and Accion Integral, for example.

In all of the above, leadership has to set priorities, and act on them. In the case of Colombia, the priority was security—the door through which much else could follow.

Still, the problem that gave rise to FARC and to social discord—a two-speed, urban–rural, centre–periphery economy of haves and have-nots—is not yet solved, not least because it is such an intractable issue, in Colombia as elsewhere. Levels of inequality may, moreover, get worse before they improve as demographic pressures grow, at least in the cities. Colombia possesses a very young population, with an additional 3 million people anticipated to enter the job market by 2025. Yet the population in rural areas is aging as young people seek opportunities away from an agricultural economy that has, by the standards of nearby Brazil and Peru, been slow to reform.

Post-settlement policies around land reform and improved land use (only an estimated 22 per cent of arable land is under cultivation) could help, both politically and economically, as a means of working towards greater inclusion and wealth. Land reform is thus not a minor problem, but rather one 'at the heart of the Colombian conflict'.[71] There are other sources and causes of violence in Colombia, but reaching an agreement with FARC on rural development will help to address the causes of conflict at the deepest level possible. Indeed, victory will be determined, in part, by the ability to remove the reasons for the emergence of conflict in the first instance, and the legitimacy these issues conferred on the guerrillas. Regardless, there are formidable constraints to be faced, from the development of rural infrastructure to the need to modernise agriculture, the importance of tax reform and ending the culture of entitlement and evasion, and the ubiquitous and seemingly endless need for improvements in education and skills. All this will take time, money and focus—and lots of it.

Aníbal Gaviria, who was governor of Antioquia for four years from January 2004, before being elected mayor of Medellín in 2012, stresses the importance of continuity of policy. 'Governments are interested in delivering big projects during their period in office, but the best things they can do are those that won't finish: health, education and infrastructure'.[72]

While Colombia's relatively high rate of urbanisation, at over 75 per cent, offers economies of scale in the country's urbanised, developed centre in delivering development, the origins of FARC as a peasant army create a political imperative for engagement in the rural areas, without which the territorial logic of stalemate cannot be broken.

The peace process itself will inevitably throw up all manner of new challenges, not least the extra-budgetary cost of inclusion of the guerrillas and the impact and cost of land reform. And in creating a new, successful model of rural development the government and FARC alike will have to confront other demons and realities, not least the fact that the image of the rural peasant working his smallholding—seemingly the aim of land reform—is out of sync with economies of scale demanded by much of modern commercial agriculture.

Meanwhile, backfilling behind military progress will never be easy for civilian agencies operating in far-flung areas across a large and topographically challenging territory of more than 1 million square kilometres. Success will rely, in particular, on getting roads and power into difficult-to-access areas, along with services and opportunities, from Internet access to jobs. Achieving this will be difficult in and of itself, given market (in)accessibility. It will also be difficult financially, given the limited fiscal latitude of the state, and politically, in terms of convincing citizens of Bogotá, Cali, Medellín and other cities—for whom Colombia's periphery is too often out-of-sight and out-of-mind—of the security and governance imperative to resolve conflicts in these areas. None of this is likely to be helped by fresh political pressures on expenditure given the inclusion of more populist-minded politicians in formal government processes.

The entry of FARC into the formal political space could serve to legitimise the role of the Colombian left, and increase its support in the process. In *Short Walks from Bogotá*, Tom Feiling cites Nicolas, a FARC guerrilla, in his description of this transformation: 'The mayor of Bogotá, the second most powerful man in the country, is from the left and he came to power without firing a single shot. We don't have to go the painful way—of armed struggle—any more.'[73]

Indeed, the peace deal creates a variety of fresh challenges, from the Bolivarianisation of Colombian politics, to the maintenance of both funding for and the integrity of the armed forces, and payment for the various aspects of the peace agreement. The three aspects that, as noted above, have characterised Colombian success since 2000—political will, technical capability and sophistication, and budgetary wherewithal—could, in the process, be threatened. A peace deal will have to take these challenges into account, particularly the danger of politics driving a process that is fiscally unsupportable.

Much rests on a change of attitude. Dealing with fiscal constraints demands greater tax efficiency, and that requires sounder administration and a different attitude on the part of Colombians towards the state. While the tax-to-GDP ratio is very low, the statutory rates are very high, with the burden falling on businesses. For example, the ratio of tax paid between businesses and individuals in developed OECD countries is 3:1; in Colombia it is 8:1. Tax avoidance has thus, since the 1980s, become a national sport. And there are other aspects of entitlement that need addressing, including urban families locked into a cycle of criminality, and the dependency of some *campesinos* on the insurgency as a means of income.

The impact of overall economic conditions and levels of opportunity on the insurgency and on criminality can, to an extent, be gauged by the state of the drug trade. According to a 2015 White House report,[74] as denoted in the table below, coca cultivation in Colombia rose 39 per cent during 2014, resulting in an increase in drug production from 185 tons to 245 tons. In part, this increase has been linked to FARC's control of the drug trade, where coca is grown under its control and patronage, and to the government's suspension of aerial eradication.[75] It is also, however, related to broader market conditions, especially concerning the price of gold. Indeed, when the gold price reached record levels of over $1,900 an ounce in 2010, many casual labourers harvesting coca (known as *raspachines*) moved across into the informal mining sector. With the fall in the gold price to under $1,200 an ounce, and the exhaustion of alluvial deposits, however, labour moved back into coca.

13. Colombia's Coca Cultivation and Production[76]

	2001	2003	2005	2007	2009	2011	2013	2014
Cultivation (Hectares)	169,800	113,850	144,000	167,000	116,000	83,000	80,500	112,000
Production Potential (Metric Tons)	700	445	500	450	260	180	185	245
Eradication from Air (Hectares)	84,251	127,112	134,474	148,435	101,573	103,302	47,052	55,553
Manual Eradication (Hectares)	1,745	4,220	37,540	64,979	60,954	34,592	22,120	11,703

Source: US Department of State, International Narcotics Control Strategy Reports.

Lessons from elsewhere?

While Colombia contains positive lessons for other countries in transition, there are lessons, too, that flow in the opposite direction. There are transition pitfalls, some of which are evident in the negative experiences of South Africa, a country with extraordinary similarity to Colombia. Such parallels include a similar size population and territory, dependence on commodity exports, a sadly comparable homicide rate (31 murders per 100,000 people for South Africa, 30.8 for Colombia),[77] a difficult regional neighbourhood, a number of major companies (SAB Miller and AngloGold Ashanti for example) and, hopefully, the end of a long-term insurgency though negotiation.

Of course, there are differences, not least the fact that Colombia does not have a racially based political system, and that those insurgents now coming into formal politics through negotiation in Colombia do not represent the popular majority.

The post-conflict experience of South Africa's armed forces should, however, flag some danger signals. This experience is linked to a radical reduction in defence expenditure since the end of apartheid, conceived initially as a means of emasculating the country's military, and subsequently as an outcome of the much-anticipated peace dividend. This has led to a South African National Defence Force (SANDF) that is chronically under-funded, over-staffed, lacking in core combat capacity, and able to deploy less than an operational brigade from its 78,000-strong armed forces. Instead of a preferred expenditure ratio of 40:30:30 on personnel, capital projects and operations, the SANDF now spends 56–58 per cent of its budget on personnel and struggles to spend between 8–12 per cent on capital projects. It is no longer a sustainable system, with the armed forces living off their reserves. As a result, according to the *South African Defence Review 2014*,[78] the government's own review, the SANDF is 'in a critical state of decline' characterised by 'force imbalance between capabilities, bloc obsolescence and unaffordability of many of its main operating systems'. Without immediate remedies, this decline will 'severely compromise' South Africa's military capabilities, the review warns.

A similar situation elsewhere has seen many other African militaries grow badly under-resourced, the effects of which can be seen in the underwhelming fighting capabilities of the Nigerian military against Boko Haram, for example, or the catastrophic display by the Malian army against Al Qaeda in the Islamic Maghreb (AQIM) and various Tuareg factions. Most European militaries view a 38 per cent spend on personnel to be in the danger zone. Many African militaries would consider 45 per cent to be the danger ceiling. In 2015, Colombia's personnel spending, by contrast, comprised 65 per cent of the total defence budget, with capital expenditure at just 6 per cent.

The reason for this breakdown relates in part to the long-term health costs of Colombia's personnel, active and retired, and their families. Yet, without adroit reorganisation and political manoeuvring, financial pressures like those experienced in post-apartheid South Africa might be envisaged for the post-conflict Colombian armed forces, as the justification for continued expenditure changes and spending priorities shift from security to social issues.

In all of this, there are inevitable challenges in maintaining growth in an economy that remains vulnerable to commodity price fluctuations and, in particular, a declining store of oil reserves.

Conclusion: Working under the iceberg

Much has been achieved in Colombia since 2000, continuing the historically conservative and sound tradition of economic management. But deep challenges remain, especially in boosting productivity, maintaining stability and reducing income inequality. In this, there is a constant imperative to diversify the economy by keeping focused on improving infrastructure, raising standards of education and health-care, and maintaining security.

Now that the threat to Colombia's cities has been relieved and FARC's leadership has been severely damaged, there is a danger that the country's elites will again switch off to the demands and concerns of rural Colombians once the immediate security threat has dissipated, unconvinced about the urgency of sharing privilege. Moreover, it has proven difficult, in Colombia as in developing counties elsewhere, to build an inclusive economic system with full

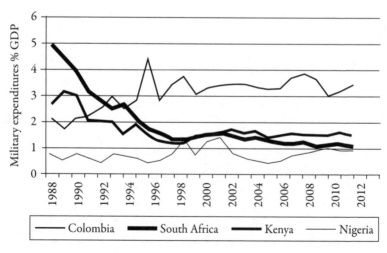

14. Comparative Trends in Defence Expenditure

employment in the short-term, not least given a generational lag in the effect of improved educational opportunities. The alternative of redistribution through welfare, meanwhile, remains fraught with long-term problems of fiscal load and a pervasive culture of elite and even guerrilla entitlement.

In this regard, the path of economic policy reform resembles the tip of a much larger 'stabilisation iceberg'. A lot has to happen underneath to create the tip that appears on top, including the provision of social services, a change in attitude from elite prosperity towards more inclusive growth, and corresponding shifts in societal attitudes towards both work and crime. Again, this has resonance in other post-conflict transitions.

In sum, social peace is inevitably a long haul. Success in this goal requires giant dollops of political will and adroit allocation of resources. Just as security is the door through which economic development follows, the task of nation-building and social inclusion will be made much easier by consistent and higher rates of economic growth.

5

FARC'S TRANSFORMATION

THE COMBINATION OF ALL FORMS OF STRUGGLE

David Spencer

What might FARC take from the peace talks in Havana? Will the guerrillas demobilise and, like M-19 before them, become one of the several political parties competing for power within the framework of institutional democracy? Or will they attempt to become something else? This chapter builds on the historical and contemporary discussion in the previous chapters with a detailed account of FARC's conceptual approach—the 'combination of all forms of struggle'—and then looks forward, considering how the peace process may influence FARC's development into the future, and how this might, in turn, affect Colombia.

Talking peace

As of May 2015, the Colombian government was negotiating the final two (victims' compensation and ending the armed conflict) of five pre-agreed topic areas for the peace accords with FARC. Optimists say that, after three previous failed efforts in 1983, 1991 and 1999, Colombia has never been this close to signing a peace deal with FARC, and that a final agreement is merely a matter of time. Once signed, according to this view, a new era will begin in Colombia.

The optimists are not so naïve, however, as to believe that all of Colombia's security problems will be solved with the peace accords. Most agree that illicit markets, violent criminal organisations and other forms of criminal enterprise will continue to plague the country. Additionally, it is unclear if and when the ELN guerrilla group—which still has several thousand fighters in the field and is a significant threat to peace—will participate in a peace process. For now, as a result, warfare between that group and the state will continue.

However, with FARC off the battlefield, one of the current major sources of conflict in Colombia will be removed. In theory, at least, this will allow economic development to occur in areas formerly dominated by FARC, where it had previously been held back by the threat of violence and extortion. If this theory proves correct, the absence of fighting will generate a virtuous cycle of greater peace leading to greater prosperity (discussed in the last chapter), and this in turn should consolidate that peace. This is what virtually all Colombians hope for.

Nevertheless, even the most optimistic understand that, should all go well, the work required to integrate former combatants and victimised populations and remove the causes that continue to propel Colombians into the ranks of violent political and criminal organisations will still take years—at minimum a decade. All this is based on the assumption that FARC's intention is to give up its desire to take power through armed struggle, turn in its weapons and be reincorporated into the existing democratic society. Even if this is true, and even under the very best of circumstances, it will not be an easy task. Moreover, as discussed in the previous chapter, it is not clear that those conditions exist; if they do not, it might be depressingly concluded that it is almost easier to revert to war than to continue to attempt to build peace.

Some among the pro-peace-process camp take a more cynical approach, believing that the accords with FARC will produce results similar to those of the 1990 demobilisation of M-19. They posit that popular support for FARC is scarce (less than 2 per cent of the population, due in part to the urban–rural population distribution discussed in Chapter 3) and fragile, and that the guerrillas' political acumen is primitive, bordering on naïve.[1] As one cynic who wishes to remain anonymous put it, 'the prize for M-19 was the 1991 Constitution, but after that they self-destructed into nothing. The same thing is going to happen to FARC.'[2]

For a number of reasons, however, we believe that the FARC peace process will not be anything like the demobilisation of M-19. There are, we believe, five important differences. First, FARC has a very well defined strategy. Second, as described in Chapter 3, FARC guerrillas are extremely well organised compared to the looser structure of M-19, which followed an urban cell-based approach. Third, FARC is highly disciplined. Fourth, the regional situation is completely different: when M-19 demobilised at the end of the Cold War, virtually all Latin American countries were becoming institutional democracies based on nominally free-market economies, and both left- and right-wing authoritarian governments had fallen out of favour. Today, however, the region is almost completely dominated by leftist and increasingly authoritarian populist governments, whose leaders share great affinities with the Bolivarian ideology of FARC.

Fifth and finally, FARC is one of the wealthiest insurgent groups in history, with coffers filled with money from drug trafficking, illegal mining, extortion and other illegal enterprises—FARC annual income is estimated at around $600 million dollars per year, a fantastic sum by any standard.[3] M-19 was far more impoverished. All this suggests to us that the potential demobilisation of FARC has very little in common—except some superficial similarities—with the demobilisation of M-19.

FARC is no Western government

It is also worth remembering that FARC does not think in the same way as liberal-democratic nation states. A perennial mistake of Colombian administrations when dealing with FARC is the tendency to engage in 'mirror imaging'—that is, consciously or unconsciously imposing one's own way of thinking on an adversary. Although FARC is composed of human beings subject to normal frailties and vicissitudes, FARC as an organisation has a number of characteristics that are misunderstood by most analysts schooled in the politics of democratic nation states.

For one, FARC's long-term strategic vision is not subject to change through periodic elections. It does not change with the latest popular political ideas; in fact, FARC is disdainful of concessions made to take advantage of temporary situations, regarding this as losing sight of strategic objectives. Rather, FARC strategy is communicated through conferences held every five to ten years. Specific programmes, modifications or adjustments are communicated through plenums, which are held as needed. An interesting characteristic of

FARC strategic thinking is that a great deal of experimentation usually occurs before a programme is officially sanctioned as a strategic line of action, sometimes two to five years in advance. For example, as described in Chapter 2, FARC's strategic approach for its mobile warfare offensive against Bogota was developed in secret, a full five years before it was even communicated to higher-level cadres, let alone publicly announced to the wider organisation.

Likewise, FARC's strategic plans and programmes often include projected dates, but meeting a date is far less important to FARC than accomplishing the specifics of the plan. In other words, to FARC, the plan will take as long as it takes, particular circumstances be damned. FARC has been very transparent about this in the Havana negotiations, reiterating this principle repeatedly. Still, many analysts among the optimists have preferred to view this as mere posturing.[4]

Second, all of FARC's leaders go through an extensive education process before joining the decision-making elite. This imbues them with homogenous doctrine, discipline and an organisational culture that values collective decision-making over strong individuals. While serious disagreements do occur within FARC, and are hotly debated in the various forums available, once a decision is made, everyone falls completely in line. Because of this, and despite setbacks, losses and desertions, FARC has been able to maintain cohesion, and leaders (even when physically separated from subordinates) have exercised command and control over the entire organisation. The notion that there are significant rogue elements within FARC is not borne out by the guerrillas' history. While there have been dissident groups, including much of the M-19 leadership and the infamous Ricardo Franco Front—which split from FARC in the 1980s and engaged in several horrendous massacres in southern Cauca—the point is that these dissidents had actually to leave FARC in order to dissent. Sometimes, especially in Colombia, the enemy of my enemy is not a friend—just another enemy.

FARC's conceptual approach

FARC's ideological foundation is not the Castroist doctrine of the Cuban Revolution, as it is for many of Latin America's prominent insurgent groups. Nor does its military theory conform to Che Guevara's notion of the revolutionary *foco* or small vanguard group. Rather, FARC's political doctrine came from the pro-Soviet Communist Party of Colombia (PCC), and its guerrilla doctrine from Mao's Protracted People's War—as heavily modified by the

Vietnamese, who were considered acceptable to the pro-Soviet FARC at a time when the Chinese were seen as heretics during the Sino–Soviet split.

As discussed in previous chapters, this has given rise to a heterodox and specifically Colombian set of ideas, within which FARC's goal is taking power through what its leaders call the 'combination of all forms of struggle'. This methodology has a long historical pedigree: it was first adopted by the Colombian Communist Party's Ninth Conference in June 1961 and ratified in the Thirtieth Plenum of the Communist Party Central Committee of June 1964.[5] FARC has made constant reference to it ever since. It is important to understand the meaning of this term, which is FARC's translation of the Vietnamese concept of 'the war of interlocking', first expounded in 1947 by Truong Chinh, a Vietnamese ideologue and student of Mao, who perfected the idea from his studies of the Chinese Civil War. The concept posits that many forms of struggle are required to achieve revolution, including the political struggle, the military struggle, the economic struggle, the propaganda struggle and many other major and sub-component forms of struggle.

For example, the military struggle is usually divided into four modes: terrorism, guerrilla warfare, mobile warfare and positional warfare. Within the political struggle, there are subdivisions of legal and illegal political struggle. An example of the former is participation in elections, while an instance of the latter is violent social protests. While pre-coordinated peaceful political protests are a recognised and legitimate form of political participation in most democracies, when these protests are spontaneous, violent and disorderly they are considered illegitimate. It is the latter that FARC refers to as 'illegal' political struggle.

The FARC concept posits that all of these forms of struggle occur simultaneously, as interlocking parts of an overall strategy, and none are discarded. However, the form of struggle that predominates during any given period, and the intensity of that effort, depends on the correlation of revolutionary and government forces at a given time and place. Over time, as this correlation shifts based on the relative strength and weakness of both the insurgents and their adversaries, in theory the mix and intensity of each of the lines of struggle changes as well.

The art of insurgent leadership, then, is the ability first to constantly and correctly analyse the correlation of forces between the guerrillas and the government, and choose the methods that will maximise the effectiveness of and minimise the risks to the insurgent force. Second, it is the ability to create an organisation that is flexible enough to constantly adapt and change in response to the shifting correlation of forces. The difference between successful and failed insurgencies lies in the ability to adapt and change with the constant adjustments of the relative correlation of forces.

These adaptations follow fairly predictable patterns. When insurgencies are stronger and governments are weaker, insurgencies adopt concentrated and more open forms of struggle. In the military arena, formations tend to be more regular, uniformed and well armed. When insurgencies are weaker and governments stronger, the insurgency becomes more dispersed and irregular. At their weakest the insurgents become clandestine organisations that rely almost exclusively on terrorism (in the sense of urban asymmetric attacks) to send their political message.

Insurgencies justify the use of violence by claiming that they have been shut off from all legal possibilities to participate in politics, and have no other options. Violence in these circumstances is the method by which insurgencies open up political space for mass mobilisation and the consolidation of their victories. Theoretically, insurgencies consolidate those spaces with political cadres and popular organisation. However, historically many insurgencies become over-focused on the use of violence—the military struggle—and cannot break out of this paradigm. It is not necessarily that the situation constrains them, but rather that insurgencies invest so much in the development of violence that it becomes mentally and physically difficult to switch gears. History is full of insurgencies that have been defeated because they could not break out of the purely military sphere of action. Indeed, the fact that insurgent victory is relatively rare—historically, guerrillas have only succeeded in overthrowing governments in less than one in five cases—is a good indication of the difficulty of leading an insurgency.

For many years, this is exactly what happened to FARC. Although FARC's leaders had always espoused the idea of the combination of all forms of struggle, it was also difficult for them to implement the concept. This was largely the result of the historical love–hate relationship between FARC and the Partido Comunista de Colombia (Colombian Communist Party, PCC). The argument mostly revolved around which organisation was subordinate to the other. The PCC always viewed FARC as one of many tools for political advancement. FARC leaders, on the other hand, clung to the notion that the PCC was at war with the Colombian state and, under the paradigm they called 'war communism', military considerations predominated over all others. This difference of opinion led to some severe clashes between FARC and the PCC. In FARC's view the PCC had forgotten its strategic goals and principles: its leaders were too willing to make strategic sacrifices to gain short-term tactical political advantages that gained them nothing in the long run. In particular, FARC saw PCC leaders as willing to sacrifice their armed wing (that is, FARC) in exchange for crumbs offered by bourgeois democracy rather than holding out for the structural changes necessary to solve the fundamental socio-political problems of Colombia. In FARC's view, there can be no peace without socialism because it is the absence of socialism that generates the conditions—discussed in detail in the last two chapters—that lead to social violence.[6]

To the PCC, FARC was too willing to involve the party in its war plans and thus expose it to attack and retaliation. This was exemplified, in its most horrific manifestation, by what happened to the Unión Patriótica (Patriotic Union, UP)[7] in the 1980s when (mostly unarmed) party members, taking part in an overt and legal political process, were murdered in huge numbers by anti-FARC drug traffickers and paramilitaries. Officially, as explained in Chapter 1, just over 1,000 were killed; unofficial claims range from 3,000 to 5,000.[8] Eventually this led to a definitive split between the PCC and FARC, and to FARC becoming what the distinguished counterinsurgency analyst Tom Marks has characterised as 'a large *foco* in search of a mass base'.[9]

Yet even as FARC focused on military growth, it still made attempts—feeble as they were—to establish a mass base. In the early to mid-1990s, for example, FARC experimented with the concept of 'co-government'. Under this concept, FARC did not need a political party, rather the guerrillas recruited sympathisers from within the existing political establishment, no matter the political banner under which they operated. These recruits carried out and supported FARC's agenda within their own organisations without being known to be members of FARC, in the manner of classic Bolshevik crypto-communist 'fractions' in mass organisations.[10] FARC leaders would meet with these elements regularly to provide orientation and guidelines. Often, they demanded that a percentage of local government budgets either be diverted directly to FARC, or used to build infrastructure or run projects that favoured FARC interests. They also used manipulation, coercion and even murder to prevent other candidates from competing against their favourites, and to 'inspire' the local population to vote for the right candidates—an approach that cost them any claim to victim status during the suppression of the UP, conducted using the same methods FARC had itself pioneered against its own opponents.

Co-government only worked as long as FARC had significant presence and armed influence in the areas in question. Once the guerrillas were displaced by government forces, most of FARC's clandestine recruits proved to be merely opportunists, switching loyalties as soon as they were no longer shackled by or benefiting from FARC.

Realising the need for a more formal arrangement, in 2000 FARC built on the co-government concept by creating the Partido Communista Clandestino de Colombia (Clandestine Colombian Communist Party, PC3), discussed in Chapter 4, as a key component of its underground political structure. The PC3 provided the virtues of co-government while avoiding its vices. People were recruited clandestinely from among normal institutions and parties, but remained active in their institutions of origin. In this way, they remained hidden from public view and, just as importantly, protected from retaliation—the importance of which was a lesson from the UP massacre. The relationship between FARC and its clandestine cadre was formalised through periods of training and indoctrination. Recruits became formal FARC cadres through the PC3; organised into classic cell structures, they became the eyes and ears of FARC within the establishment. Moreover, unlike co-government recruits, when left alone, members of the PC3 have remained loyal to FARC. From time to time these operatives have been discovered, sometimes in very high positions of public trust or influence.[11]

However, understanding that not all PC3 cadres would be equally ideologically committed to the cause, and that many others could share some or all of FARC's goals without being members of either FARC or the PC3, FARC created another organisation, in which these fellow travellers could participate: the Movimiento Bolivariano por la Nueva Colombia (Bolivarian Movement for the New Colombia, MBNC). The vanguard of this movement was to be the PC3, though information as to who was a member of the Bolivarian Movement and who was PC3 would not be revealed. The Bolivarian Movement was launched with much local fanfare in the El Caguán demilitarised zone in April 2000, under the direction of then-FARC Secretariat member Alfonso Cano. Yet to the broader Colombian public at the time, both MBNC and PC3 seemed to be almost stillborn. Popular opinion was overwhelmingly against FARC, and even radical leftist parties, natural candidates for entry into the MBNC, seemed to shun it.

Upping the military ante

Meanwhile, under the direction of its founding leader Manuel Marulanda, FARC increasingly raised the military stakes, attempting to achieve a definitive breakthrough by means of the military struggle. However, government forces proved to be more adaptable than FARC. Where FARC remained militarily rigid and inflexible, trying to achieve a breakthrough by doing more of the same, the military (as explained in Chapter 2) adopted new and innovative tactics and methods, and leveraged external assistance to build transformative new capabilities. Supported by popular will, a friendly United States and eventually the political establishment under President Álvaro Uribe, the government became willing to do whatever was necessary within the law to defeat the insurgency.

As argued in Chapter 4, FARC's fundamental strategic error was to transition to a 'war of movement', expanding beyond the bounds of the rural periphery where it could always rely on local support, and trying to confront the government on its own turf, closer to the country's urbanised and developed centre.

The threat of FARC encirclement prompted a national crisis, and the military backlash against the guerrilla offensive of 1996–99 began the slow but steady rollback of FARC. First, the Colombian military disarticulated the guerrilla fronts besieging Bogotá. Then they took the war to FARC's turf, launching a deep and prolonged offensive into the insurgents' south-eastern strategic base area, denying them the ability to project power across the rest of the country. Additionally, aided by Plan Colombia, the police and army launched a sustained campaign against drug trafficking, FARC's major source of financing.

Once this programme—which came in the form of Uribe's Democratic Security Policy, discussed in the Introduction and Chapter 1—was fully implemented by the beginning of 2004, the government rapidly recovered those areas of the country that had long been under FARC's sway and all indicators of criminality (murders, kidnappings and so forth) began to drop dramatically. When FARC tried to gain tactical victories by massing large numbers of guerrillas, the formations were detected and decimated. FARC were forced to break up into small guerrilla groups and fight a war of attrition against the security forces, relying largely on homemade improvised explosives to slow the army's progress. By late 2007, as explained in Chapter 2, the Colombian Air Force had acquired smart bomb technology in the form of the US Paveway II system, and began to carry out a series of spectacular raids that killed high- and mid-level FARC commanders at an unprecedented rate. In March 2008 alone, three FARC Secretariat members died—Raúl Reyes, Iván Ríos and Manuel Marulanda. Then-commander of the Colombian military forces, General Freddy Padilla, began talking about the 'end of the end' of FARC.[12] It seemed he might be right when government forces were able to fool FARC and carry out the spectacular rescue of fifteen high-profile hostages (including three Americans) in Operation Jaque (described in Chapter 2), an achievement that was enabled by the scattering and isolation of FARC units in previous operations. However, the Colombian government's talk of victory was premature.

Within five months of replacing Manuel Marulanda in 2008, Alfonso Cano, the new FARC general secretary, outlined his strategic thinking in a letter to the rest of the Secretariat. This letter, as we saw in Chapter 1, became known as the 'Plan Renacer Revolucionario de las Masas' ('Plan Revolutionary Rebirth of the Masses'). Most analysts concentrated on the military aspects of the plan, but its title accurately reflected Cano's priorities, which were essentially political. His letter contained fourteen numbered points, and his very first point was the need to share the plan with the Venezuelans and with other political allies of the Colombian left to help them create a 'people's party' and to develop an alliance between this party and the continental Bolivarian Movement.[13] His second point was that where FARC had lost the most ground was in the political-social arena, and that the key to success was the 'revolutionary rebirth of the masses'.[14] His remaining twelve points were about the implementation of guerrilla warfare to enable this rebirth of the masses. Although not stated explicitly, it is clear from the context and tone of the document that Cano had no intention of abandoning FARC's goal of taking power, but rather planned to adapt to the Colombian government's shifting approach and to change the particular mix of 'all forms of struggle' to achieve FARC's unchanged strategic objectives.

That his thoughts were revolving around this change of the mix of the forms of struggle was confirmed by many of the documents captured when Cano was killed on 4 November 2011 in Operation Odiseo. One was a well-marked edition of the Spanish-language translation of Gene Sharp's *From Dictatorship to Democracy*, which can be downloaded for free from the Internet.[15] Sharp's book is about mobilising people to carry out non-violent civil disobedience and protest to overthrow a dictatorship, and Cano was clearly considering Sharp's ideas. Yet there are two key problems in applying Sharp's thesis in Colombia. First, the government is not a dictatorship by any stretch of the imagination, and yet FARC apparently thought it needed to be overthrown through popular mobilisation. Second, under Cano and subsequently Timochenko, FARC showed very little sign of becoming non-violent.

In the peace talks, FARC negotiators have spoken about agreeing to stop using weapons, but not about giving up their weapons.

The most important example of the approach Cano seems to have been considering at this time is not found in Gene Sharp's ideas, however, but in nearby Bolivia. Most readers probably do not view the 2005 electoral triumph of Evo Morales and his Movimiento al Socialismo (Movement toward Socialism, MAS) party as the product of an insurgency, but it was exactly this—a brilliantly conceived and executed application of the concept of the combination of all forms of struggle to achieve maximum advantage within a correctly determined correlation-of-forces analysis.

The Bolivian model was composed of five key elements:

Violence

Violence was limited in the Bolivian case, but not unimportant. It was employed judiciously in two principal ways. First, guerrilla warfare was used to protect the Cocaleros'[16] main source of income (coca production) and was strictly limited to drug-producing areas—that is, the Chapare region of Bolivia. Second, flying columns accompanied social protests during key events, employing selective violence to produce incidents between anti-riot forces and the crowds. The goal was to up the ante, embolden the masses and delegitimise the government. Every attempt was made to keep the use of violence 'invisible', in the sense that in the Chapare region it was attributed to individuals and amorphous 'drug traffickers' reacting to counter-drug operations, while during the protests it was presented as marchers reacting to government suppression, rather than as a deliberate or systematic line of effort by the Cocaleros. The Cocaleros and their allies were portrayed as innocents exercising their constitutional rights. This limited and selective use of violence made it politically difficult, even undesirable (because it seemed such an extreme overreaction), for the government to denounce the Cocaleros as insurgents. This played into Cocalero hands because it meant that successive administrations failed to expose the real danger of the Cocaleros and their intention to overthrow the Bolivian constitutional order. Towards the end (2002–4), when they were near final victory, the Cocaleros were less circumspect about revealing their armed might, and it was this threat of violence rather than violence itself that caused the government to make the final concessions that led to Morales's election.

Drug money

Second, in Bolivia money gathered from quotas on coca production and trafficking was used to finance the violence, the social protests and eventually the creation, expansion and activities of the Cocaleros' political party, the MAS. It was also used to buy off key opposition figures or groups. The steady flow of money was power: it gave the Cocaleros cohesion, staying power and the ability to manipulate their environment.

Manipulation of social protest

Third, social protests were the main 'visible' effort of Morales's insurgency. These protests were neither random nor spontaneous. Each sub-component of the Cocalero unions, each known as a *senda* (trail), was assigned a quota of marchers that were to participate in the protests for a predetermined period of time (usually eighteen months) during which that *senda* individual, was known as being *en comision* (commissioned). The aggregate of this professional 'manoeuvre force' was led by a highly trained cadre of junior and intermediate leaders operating under a very efficient chain of command. During their 'commissions', the Federations[17] paid marchers a salary and took care of their families. This professional core of Cocalero protesters allied themselves with other social groups such as the labour unions, miners and indigenous groups. This allowed them to rapidly mobilise around just about every anti-government cause, and carry out prolonged marches, which were far more intense than their original causes merited—keeping the country in a state of perpetual crisis.

Legal, overt politics

Fourth, the MAS was formed by the Cocaleros, as a legal political party to participate in the normal political process. This party offered itself as the solution to the social chaos of its own creation, and served to legally consolidate the gains obtained via illegal and illegitimate means.

International propaganda

Fifth, a key element of the insurgency in Bolivia was an information campaign, aided by some, mostly European, NGOs, which presented the unrest not as a struggle to perpetuate the narcotics trade, but as a struggle for long-suppressed indigenous rights. Thus, confronting the Cocaleros became politically incorrect, synonymous with repressing poor indigenous peoples.

These components bear a strong resemblance to the elements of the FARC system described in our analysis of guerrilla and counter-guerrilla warfare in Chapter 3. Of course, no insurgency is a carbon copy of another, but (as we explain below) there is evidence to support the idea that a similar mix of factors—that is, a similar combination of all forms of struggle—lies behind FARC's current approach to the peace talks in Colombia.

FARC's peace aims

Remarkably, FARC's leaders and peace negotiators have been very transparent about what their objectives are in the peace talks, and about what they really want. The problem is that this is often not taken seriously—it is seen only as posturing and people assume that what FARC leaders want is somehow short of what they have stated, repeatedly, as their goal since the first day of the negotiations in October 2012.

Based on its own statements, FARC does not wish to surrender its arms and become incorporated into democratic society, but rather seeks to transform the Colombian state into a socialist government.[18] In other words, peace negotiations are part of a plan to take power: they are not a means to end the conflict, but rather to transform it. FARC leaders have simply changed their focus to a different mix of the combination of all forms of struggle, in order to achieve that power. In recognition of this, it is useful to compare the consistent themes that have emerged in the negotiations so far with the five components of the insurgent model developed by the MAS in Bolivia, as discussed in the previous section.

Violence

The most notable aspect of the FARC negotiations is that FARC leaders have steadfastly insisted that they will not turn over their weapons and will not demobilise as an organisation. In other words, the guerrillas will not give up their latent ability to use violence in pursuit of their political objectives. If the government fails to comply with the accords, FARC will return to armed struggle. Towards the end of the Bolivian insurgency, as noted earlier, the threat of massive violence became more important than the actual use of violence. As a means of guaranteeing Colombia's march to socialism, this latent threat of violence could be key.

Additionally, it is likely that as FARC switches from guerrilla attacks to mass protests in pursuit of what was not negotiated or conceded in Havana, selective violence will be used to enhance the impact of the social movements. The range of violent actions may vary from as little as rock throwing to military-style attacks such as ambushes and sabotage. The purpose of this violence is to stir up the crowds against the government and create a climate of media hysteria to allow concessions to be wrung from the government. Knowing what level of violence to employ is tricky, because too little fails to accomplish the aim, but too much can backfire. In Bolivia, the Cocaleros learned by experience and generally kept actions on the lower end of the scale of violence, targeting the police or military to provoke them until some element of the government reacted badly (and in a way that could be captured on film or recorded by the media), with the subsequent media frenzy 'justifying' generalised violence that would not be blamed on the protesters.

It is likely that this will not be the only use or threat of violence in Colombia, however. According to one informed source, about 40 per cent of FARC's forces will not join the political process, but are likely to opt to continue their involvement in illicit economies and drug trafficking.[19] They would, as such, become FARC criminal gangs (or FARCRIM). The total compliance by the whole of FARC with the current unilateral ceasefire to date, however, demonstrates the organisation's cohesion, command and control. For 40 per cent to suddenly break this discipline and become independent drug-trafficking gangs would seem oddly uncharacteristic. It is more likely that this would be a deliberate strategy, and while publicly they would be labelled dissidents from FARC, these FARCRIM would in fact likely continue to protect FARC's major source of income, drug trafficking.

Illicit financing

In this sense, drug trafficking and other illicit economies will likely continue to be a mainstay for FARC (or whatever name the group adopts when it transitions to 'peace'). As noted above, its illicit enterprises earn FARC approximately $600 million per year. FARC is not just going to give up that fantastic sum: rather, the 'dissidents' will provide the organisation with a deniable means to preserve that money flow.

In addition, the drug-trafficking agreement signed in May 2014 as part of the peace accords raises the spectre of a rollback on the gains of Plan Colombia. First, the spraying of drug crops disappears under this yet to be implemented agreement, with only manual eradication permitted.[20] Second, the government is required to negotiate eradication with local communities, and eradication then only takes place voluntarily after a series of social programmes and welfare benefits are delivered. It is doubtful that the Colombian government has all of the necessary funding, human and infrastructural resources to fulfil all of the preconditions required for communities to manually eradicate their coca. This means that the process risks being delayed nearly indefinitely, which in turn means coca will continue to be a huge source of income for FARC even after the peace accords are signed. The government negotiators were able to introduce a clause indicating that if the government had fulfilled its promises and the community still refused to eradicate it, the government could impose manual eradication on the community. Potentially, however, communities could squabble about this indefinitely, as well as filing legal injunctions to prevent forcible eradication of their coca, ultimately making it difficult to stem the illicit financing of FARC.[21]

Manipulation of social protest

Large-scale protest action is likely to become the main form of struggle after the negotiations. Sources with access to FARC indicate that military commanders are currently being replaced with political commanders. This would mean theoretically that the guerrillas are preparing to transition from military operations to political action. However, the main political effort will probably not be what could be called 'normal' politics. Such politics in a democracy is the process of electoral plurality: at the basic level, voting for candidates representing different political parties, and at more advanced levels, participating in party meetings and conventions to develop policy platforms, mobilise party recruits and choose candidates for upcoming elections. While this is one component of FARC's vision for political participation after a peace agreement, it would not be the organisation's main form of political struggle. That will be the manipulation of protest movements.

Understanding that the government will not willingly accept the transformation of the state into a socialist system in Havana, FARC negotiators have openly called for an informal means of political decision-making. They have called on social movements to go out into the streets and protest. The idea is that if there is enough social protest, this will force the government to make concessions. Just about every communiqué and every FARC commentary calls on the whole rainbow of opposition social movements—from peasant and labour organisations to gay, lesbian and transgender movements—to go out and act to create an alternative, informal voice that the government cannot ignore. They call organisations to action by quoting Simon Bolivar's saying: 'the sovereignty of the people is the only legitimate authority of nations'.

FARC believes that if it can incite the masses to march, then the 'the government will have to listen to the people'.[22] FARC has actively fomented the mobilisation of at least three major social protests during the peace process. The first major violent protest, in August and September 2013, mobilised more than 60,000 people and shut down roads and highways all over the country. Marches occurred in parts of the country, such as Boyacá, where FARC had not previously been able to mobilise. The second violent protest occurred in April 2014, just before national elections. The third was peaceful and occurred in April 2015, when FARC claimed that 300,000 marched in Bogotá alone, with several hundred thousand more across Colombia.[23] True or not, FARC has proven that it is able to mobilise a significant number of protesters, and either provoke them to behave violently, or ensure that they behave peacefully.

How important the marches are to FARC is illustrated by the political participation agreement made in Havana. A requirement to strengthen and enhance recognition of social movements—meaning protesters—was mentioned in the second paragraph of this portion of the agreement. Furthermore, the rights of social movements and political protests took up four of the nineteen pages of the political participation document. The only other topic that took up this much space in the agreement was the issue of security measures for opposition political party members.[24]

The agreement called on the government to adopt measures to guarantee the recognition, strengthening and empowerment of social movements. It obligates the state to consult with social movements in the development or modification of its public policy. The state is to provide legal and technical support to social movements. It also has to facilitate financing of social movements and grant them access to public contracts, and to help them develop and exchange lessons learned with other social movements. Furthermore, the state has to guarantee the representation of social movements in decision-making actions at the local, municipal, departmental and national levels. The agreement obligates the government to respond in a timely fashion to petitions or proposals by social movements. While increasing democratic representation is clearly a good thing, it appears from this draft agreement that social movements are to be given rights and privileges that even political parties do not possess.

The draft agreement recognised social protest as a legitimate form of political expression and called on the government to guarantee social movements' full mobilisation, to guarantee the rights of protesters, to modify the current legal code regarding protests, and to exert greater oversight and control of authorities involved in controlling social protest, as well as obligating the Public Ministry to accompany social protests as a form of protection for protesters. Finally, the agreement called on the government to implement special protection measures for those participating in opposition politics.

While these measures are not unreasonable by themselves, especially given Colombia's history of violence against demobilised opposition forces, the problem is that there is nothing in the accords about the responsibilities of social movements to operate within the law, or to respect the rights of non-protesters. In effect, the government is expected to fund, promote and protect opposition social movements whose main function is to go out into the street and create chaos, but the social movements are not correspondingly obligated to recognise any measures to respect the peace, security and property rights of others.

Given that FARC leaders have been reading Gene Sharp, and that they rhetorically regard the government as a 'dictatorship' (or at least a 'pseudo-dictatorship') to be overthrown, it seems that FARC is asking that the illegitimate—by its definition—government promote and protect the forces that FARC intends to use to overthrow that very same government. The only explanation for why the Colombian government—after all the military successes it has achieved over the past fifteen years—appears to have agreed to these terms may be that it perhaps underestimates FARC.

Overt, legal politics

In common with the Bolivian model, and to consolidate what it gains through illicit political action and violent protest, FARC needs a recognised political mechanism to 'legalise' its actions. One of the great lessons of the Bolivian insurgency was the way in which the Cocaleros avoided taking power by force of arms. Rather, they did all but take power, and then stepped back at the last stage and legalised their victory through elections and political accords. In this manner, the great Bolivian insurgency went nearly unperceived, and although irregular, did not elicit international opposition. In our view, this is why FARC has been so insistent on characterising all of its actions (throughout its long insurgency) as merely political rebellion, and calling at the negotiations for no jail time for any of its members. Its goals include freeing those that are already in jail, as well as those who have been extradited to the US, including leaders such as Simón Trinidad, a senior FARC leader extradited to the US on murder and drug-trafficking charges in 2004.[25] To sweeten the deal, FARC has offered to issue a blanket amnesty to all members of the military accused of human rights abuses and 'false positives'.[26] This suggests the importance to FARC negotiators of whitewashing their organisation to achieve power 'cleanly'.

After legalising its acts of rebellion, the next step for FARC is to form its political party. The draft political agreement under discussion in Havana as part of the peace talks covers the formation, participation and protection of this political party. FARC has bitter memories of its last, ill-fated, attempt to form an overt political party—the UP—and in the signed agreement on political participation it noticeably seeks to avoid the mistakes of the past. However, the agreement also creates special representation for the peoples and zones especially affected by the conflict. While there is no agreement on the number of congressional representatives that will be included in this category or the number of terms they may serve, and while the draft agreement does not specify FARC in this category, these are essentially guaranteed non-competitive political positions for FARC.[27] They offer a jumping-off point for FARC representatives to begin to consolidate their political power. The Bolivian Cocaleros started out with two lower-house congressional seats, and were able to consolidate power from there, proving that it can be done.

It appears that FARC is banking on being able to woo or co-opt much of the Colombian countryside. The agreement on agrarian reform essentially socialises a large part of the Colombian agrarian space—the rural periphery into which the central state has always struggled to project its authority—and propels FARC or its political party into the role of champions and defenders of that socialisation process.[28] This would give FARC a constituency that it did not previously have. While three-quarters of Colombians live in cities, it is from the countryside that Colombia's cities get their food and supplies, and it is from this countryside that life can be made difficult for urban dwellers through protests. As the protests make greater headway, as in Bolivia, FARC's legal party, or whatever name they choose to re-brand themselves as, could potentially offer itself as a solution to social conflicts of its own making. Protests and social violence will be used to create space—and then concessions and elections will be used to legalise and consolidate that space. In Bolivia, it took the Cocaleros approximately ten years to carry out and consolidate their insurgency using this method.

It could be asked, if the government is making all of the above concessions, what do FARC and the government have to fight about? The mobilisation banner for the upcoming protests will likely be the failure of the government to fulfil the peace accords. The risk is that Colombia is repeating a historical pattern—which we characterised, in Chapter 2, as the dynamic of stalemate—in which wars were fought almost to military victory, but then concluded with a bad peace deal that set the conditions and causes for the next round of conflict. This peace process is unlikely to be any different. What is striking about all three of the partial agreements that have been signed so far—on agrarian reform, drug trafficking and political participation—is the lengthy and comprehensive lists of things, in each of these areas, that the government is required to do to. It is stunning that FARC are required to do nothing in return. It is not clear why the government is being so accommodating.

Moreover, calculation of the material, human and infrastructure costs associated with each of these items indicates that the difficulty facing the government in performing all of these tasks is likely to be overwhelming. In fact, it will probably be impossible for it to carry out all of these tasks to a satisfactory, or even to a minimal level. This is especially true when one considers the state of the Colombian economy, which (as discussed in the previous chapter) has recently seen a downturn in its pattern of sustained growth due to the global drop in the price of oil. At the same time, Colombia has not been able to mobilise the international donor support that it perhaps had hoping would support the demobilisation, disarmament and reintegration process being contemplated, probably because many potential donors are facing their own internal economic challenges. Inability to satisfy some or many of its promises may fuel the protests against the government that FARC is planning on using as its main tool for mass political mobilisation.

Additionally, while the government is contemplating a referendum to obtain popular assent for the peace agreement, FARC's leaders have a very different idea, and are calling for a new constitutional convention. While FARC agreed to limit the negotiating points to the five pre-agreed subjects, it believes that in order to consolidate the peace, it must change the entire political and economic order of Colombia.[29] Needless to say, the government's negotiators have been unwilling to give in on this point, which would amount to surrendering at the negotiating table all of the military gains achieved since 1999.

FARC's leaders have offered to participate in a partial constitutional convention, where a limited number of points of the law are debated and changed. However, this is a Pandora's box, because once the constitution is modified in one area, this will require amendments in other areas to balance the areas that were changed. Furthermore, FARC has a very fixed idea about who will be included in the constitutional convention. According to FARC leaders, the convention would include 141 delegates who would be both elected and appointed from amongst the guerrillas, peasants, indigenous communities, Afro-Colombians, the victims of the conflict, organised groups of women, students, workers, the lesbian, gay, bisexual and transgender (LGBT) community, retired military, residents, refugees and exiles, traditional economic power groups, social power groups and political parties.[30]

If FARC was to get its way on this matter, minorities would be over-represented and the traditional powers heavily under-represented, meaning that the emerging constitution would most likely favour FARC. While it is not illegitimate to ask for these groups to be represented, it would be a problem if they were over-represented. This is exactly FARC's goal, of course, in order to give itself a majority in the constitutional convention, which would then allow its representatives to bulldoze the opposition and establish a socialist constitution. On the other hand, the refusal of the government to grant such a heavily skewed constitutional convention could simply become another banner for the mobilisation of protesters by FARC and a justification for violence.

International propaganda

In the past, FARC was not very adept at public communications. But FARC leaders have proven to be surprisingly proficient in this area during the peace process. At least four websites have carried stories about the negotiations, proposals and advances to the outside world, and FARC has sponsored multiple affiliated websites, which do the same. FARC has issued a constant stream of bulletins, press conferences, stories and videos on these sites. FARC propagandists have been remarkably agile compared to the government, which has posted very little. In many ways, FARC has been more transparent and credible than the government in this regard. Virtually everything FARC has done in the area of international propaganda has been through the Internet and non-traditional media. FARC leaders have complained bitterly that they have unequal access to traditional media, believing that this form of media is biased against them and controlled by the government and its interests. They have pointed out that of 140 media outlets that have covered certain aspects of the peace negotiations, just two have been 'alternative' media outlets—in other words, those trusted by FARC. Its unhappiness with the lack of 'fair' coverage also feeds into its desire for increased mass social protest, to get its message out to the broader public.

The consistent message that FARC sends out through its 'alternative media' sources aims to discredit the government by establishing and reinforcing the perception that FARC seeks peace while the government is only interested in FARC's surrender, and that government negotiators are therefore maintaining inflexible positions towards FARC proposals. FARC outlets have furthered this perception by consistently and publicly asking the government to consider issues either outside the scope of the negotiating framework, or unacceptable to significant constituencies within the Santos government. A case in point has been the bilateral ceasefire. This would seem like a good thing, a step towards peace. Without more concrete agreements, however, this is problematic for the government, particularly in the face of other threats such as the ELN and the BACRIM.

Neutralising the security forces

A final element of FARC's approach—one that falls outside of the scope of the five-part Bolivian model—can be seen in the way that FARC has maintained its attempts to undermine the Colombian military. The armed forces have been the main obstacle to FARC's advance in its quest to take power. In response, FARC has attempted to weaken the military in two ways. First, it has sought to create the perception that the Colombian military campaign has been useless, and to paint the armed forces as warmongers. Second, FARC negotiators have proposed measures to significantly reduce and 'defang' the military.

For example, in its information war, FARC has made consistent statements to the effect that despite all of the state's investment in military capability, the Colombian armed forces have proven unable to defeat the FARC guerrillas. FARC has accused officers in favour of continuing the military campaign of searching for 'the final solution'—equating the military with the Nazis and the war with the Holocaust. FARC spread the idea that US Southern Command's campaign plan has been defeated and that despite this effort, the armed confrontation has spread intensely to all of Colombia's national territory, rather than diminishing because of US involvement. This promotes the notion that the Colombian military are anti-patriotic servants of a foreign power only capable of operating with the aid of that foreign power, and that even so, the foreign power has been defeated. Quietly, FARC leaders also admit that they have suffered heavy losses, particularly—as described in Chapter 2—in the face of Colombian air power (one of the main programmes under the

US-supported Plan Colombia), which of course is true and is one of the principal reasons that they have been so keen on weakening the military.

In the peace talks, the FARC proposal is termed 'Immediate and Preventive Actions for Non-Repetition'. FARC has proposed that in order to reconcile society, the state should renounce COIN, creating a political culture that recognises social struggle as a legitimate form of democratic expression and places a premium on human rights. In other words, FARC has called for the legitimisation of insurgency and, in so doing, the de-legitimisation of COIN. This includes reducing the military budget to 2 per cent of GDP (which, as noted in the last chapter, would represent a very significant cut to current military spending) and transforming the military to give it a mission oriented exclusively around external defence. FARC has proposed that all foreign assistance or advice be banned, has called for a reform of service schools and doctrine to cover only external defence, and has suggested a redefinition of civilian defence and security-sector jobs. Furthermore, it has called for externally focused intelligence services and proposed the separation of the police from the Ministry of Defence and its subordination to the Ministry of the Interior.[31]

Essentially, these measures would serve to undo all of the progress made by the Colombian military under Plan Colombia and Uribe's Democratic Security Policy, returning the armed forces to their pre-Plan Colombia state of the late 1990s. Beyond this, alterations to the military's doctrine and mission would prohibit the armed forces from participating in activities relating to internal defence, including their ability to control mass protests. Separating the police from the Ministry of Defence would additionally neutralise a measure that has made the Colombian security forces so effective over the last decade: inter-service cooperation and intelligence sharing. Furthermore, prohibiting foreign assistance and advice would prevent the United States from continuing to assist Colombia, even in an emergency such as a repeat of the national crisis that led to the establishment of Plan Colombia in the first place. Finally, should FARC decide to return to military action as a form of struggle, its task would be appreciably easier against this much-weakened and limited military institution.

Interestingly, FARC has also called for the military and intelligence archives to be opened to the public. All intelligence service members that have participated in 'creating victims' of the internal war are to be purged.

Finally, FARC has called for a commission on 'paramilitarism', with the goal of purging all members of the security forces accused of having links to paramilitaries.[32] These measures are justified supposedly by the fact that they would grant victims access to the truth—and perhaps in part they would. But the commission's real purpose (from FARC's standpoint) would be to identify the sources, methods and individuals that were effective in combating the organisation. This would allow FARC to develop counter-measures and could potentially put those who supported the government against FARC at serious risk. It goes without saying that anyone who acted illegitimately should be identified and punished, but those who acted within the law and in good faith should be fiercely protected.

Conclusions

For Colombia's sake, it would be wonderful if FARC's current positions in the peace talks were merely posturing. But as we have explained, such a hope is inconsistent with the organisation's history and strategic concept. It may be possible that FARC has turned over a new leaf, and that it intends to join Colombia's institutional democracy as just another political party. If this is the case, it should not be held against FARC leaders that they should seek every advantage possible, and negotiate the best deal possible to position their organisation for this endeavour. Yet, to date, FARC's proposals and petitions have resembled not those of an organisation that seeks to reincorporate itself into normal society, but rather those of one attempting to dictate at the negotiating table the terms of a peace that it was unable to win on the battlefield. Indeed, FARC's statements and proposals are more consistent with an organisation that is still trying to take power—simply by different means.

As we have suggested, there is evidence to suggest that normal democratic politics is not what FARC leaders have in mind. Rather, it appears more likely that they will form a political party and participate in elections, but that this will be backed up by massive, violent and semi-violent social protests, funded by drug trafficking or other forms of illegal activity, and protected by armed criminal bands. If, indeed, the successful Bolivian insurgency is FARC's future model, this would transform the game entirely. In effect, it would mean that the current FARC negotiating stance represents merely a change in the mix and emphasis of the combination of all forms of struggle, rather than an abandonment of armed and other illegal forms of conflict.

6

COLOMBIA IN COMPARATIVE CONTEXT

Dickie Davis

All insurgencies, and therefore all counterinsurgencies, differ from each other. Likewise, as explained in the introduction, one of the key goals of this book is to offer a detailed examination of how counterinsurgencies conducted by a national government on its own home turf may differ significantly from the more 'classical' (but far less historically common) expeditionary cases—such as Malaya, Algeria or Vietnam—in which an external intervening power seeks to conduct COIN in somebody else's country. Nevertheless, over the course of several fieldwork trips to different parts of Colombia whilst researching this book, it occurred to those of the authors with experience of Iraq, Afghanistan and other expeditionary COIN campaigns that there were a number of familiar themes coming through, despite the considerable differences.

As a middle-income country Colombia has had comparatively limited engagement with the international aid community: the last UN-coordinated appeal for Colombia was a Humanitarian Action Plan in 2003, and the influence of outsiders is relatively small.[1] But as we spoke to those engaged in the conflict, the authors often heard how military officers had studied the Malayan Emergency of 1948–60. In a way, this was unsurprising since (as

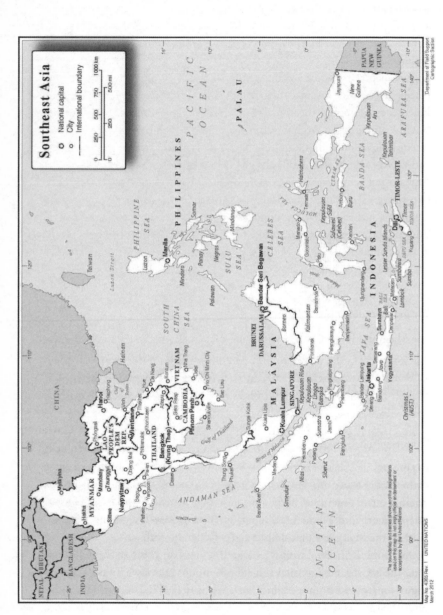

15. Map of Malaysia

Department of Field Support
Cartographic Section

described in the Introduction) that campaign has been used extensively as the baseline for much classical COIN doctrine; in another way, however, it was surprising given that so many counterinsurgencies have been conducted since. For instance, the war in El Salvador was also often quoted as an appropriate regional comparison.

This chapter, therefore sets the war in Colombia in comparative context by placing it alongside instances of expeditionary COIN including the Malayan Emergency—as noted, the source of much classical COIN doctrine—and the NATO campaign in Afghanistan since 2001, which has prompted much recent innovation and questioning of the traditional wisdom on insurgency and COIN. Three of the four authors of this book have field experience in Afghanistan, while one served on counterinsurgency duties in Malaysia, and two previously studied the Malayan Emergency in detail. By examining Colombia against the background of these very different conflicts, the goal of this chapter is to draw out the key themes that make Colombia—as an instance of internal COIN—both similar to, and yet distinct from, examples of classical expeditionary COIN.

The Malayan Emergency, 1948–60

The Malayan Emergency is best considered in four phases: the origins of the insurgency and its beginning from 1947–50; the Briggs Plan of 1950–52; unity of civil military command under General Sir Gerald Templer, 1952–54; and finally the tailing off of the campaign in the period 1954–60, although it can be argued that it did not come to a complete end until 1989.

Origins of the insurgency

In a process that started in the late 1700s, the nine states and two settlements of the Malay Peninsula first became British protectorates and then joined together into a Federation of Malaya, with some states retaining considerable independence under indirect colonial rule by British Residents and Advisers. In 1929, the Malayan Communist Party (MCP) was formed with the aim of removing the British colonial administration and establishing a communist state. The MCP gained considerable support from Malayans of ethnic Chinese origin—many of whom were in Malaya as plantation labourers, miners or (after 1931) as refugees from conflict in mainland China. At the same time, a very large number of Indian labourers—particularly Tamils and Bengalis—had settled in Malaya, and they too wanted an end to British rule. Along with marginalised and politically excluded labourers, Malayan Chinese squatters, living in illegal settlements strung out on the jungle fringe, became a key source of support for the communists. By 1937, the MCP was strong enough to organise a wave of strikes aimed at damaging the economy. The majority of the Malayan economy depended upon British-run rubber plantations and tin mines. Their output was such that they were an important source of income for Britain's colonial empire as a whole.

The Second World War disrupted matters considerably. Initially, the MCP continued to foment labour unrest in a bid to hamper the British war effort, but with the entry of the Soviet Union into the war on the Allied side in 1941, Moscow directed Comintern-controlled fraternal communist parties such as the MCP to cooperate with Allied governments. Combined with the threat of a Japanese invasion of the Malayan Peninsula—which became acute following the Japanese occupation of French Indochina in June 1941 after the outbreak of war with the Soviet Union—the MCP and the British joined forces in an uneasy alliance. This gave birth to the Malayan Peoples' Anti-Japanese Army (MPAJA), which was trained and supported by the British and, throughout the war, received training and weapons support from the British Special Operations Executive, its Australian equivalent, Special Operations Australia, and the Allied Intelligence Bureau. Ironically, Chin Peng, who became leader of the MCP in 1948, was made an officer of the Order of the British Empire for his actions as a fighter in the MPAJA. The sudden surrender of Japanese forces in August 1945 meant that Malaya avoided being fought over a second time and the MPAJA survived the war in good shape with thousands of trained cadres and carefully cached weapons that they did not surrender after 1945.

The period immediately following the war was chaotic. Commonwealth forces arrived to take control and support the police force, which was in poor shape and in need of considerable reorganisation. A demobilisation programme for resistance fighters was established, but a hard-core of approximately 4,000 MPAJA fighters went into hiding ready to support the MCP if needed. The period 1946–48 saw the British Colonial Office create a plan for a Malayan Union, which united the four Federated Malay States, five Unfederated States and two Straits Settlements that had existed before the war, leaving Singapore as a separate colony. This plan required the sultans of the Malay States to cede their authority to the Crown. This inflamed tensions and the Colonial Office quickly realised their mistake. The resulting compromise was the creation of a new Federation of Malaya headed by a British high commissioner, but with Malay sultans remaining in power in the states, and Singapore remaining as a separate Crown colony under direct British rule. The new Federation came into being on 1 February 1948. This new governance arrangement was perceived as biased towards ethnic Malays, uniting the large ethnic Chinese and Indian communities against it, something that was exploited by the MCP. The fact that many Malays (and some Indians) had collaborated with the Japanese occupation forces—while many ethnic Chinese had opposed them through the MPAJA—further exacerbated ethnic tensions.

Immediately following the cessation of hostilities in 1945, the MCP lacked the policies to exploit its strength, but between 1946 and 1947 the party resumed infiltration of the unions to create industrial unrest, and opposed the creation of a Federation. In February 1948, the MCP, encouraged by the Soviet Union, went on the offensive and sparked a wave of unrest and violence. It also started to organise a Malayan People's Anti-British Army, which later became the Malayan Races Liberation Army (MRLA) as the MCP attempted (with very limited success) to expand its support base beyond the Chinese community. The guerrillas' plan was to drive the state out of isolated areas, establishing liberated zones, and then link the zones together, thus winning control of the country. The state responded with a series of restrictive measures and the MCP campaign in turn became more violent, resulting in the murder of three European rubber-plant managers in mid-June 1948. This led to the declaration of the state of emergency on 17 June 1948, a little earlier than the MCP had anticipated. The majority of ethnic Malays and Indians had little support for Chinese calls for the creation of a communist state. Thus it was the battle for control of the hearts and minds of the ethnic Chinese civilians in Malaya that became the centre stage for the COIN campaign.

Sir Henry Gurney became the British high commissioner to the Malayan Federation on 1 October 1948, shortly after the state of emergency had been declared. Under the new federated arrangements, the Malayan states were reasonably independent of the federal government; the Malayan Civil Service was small and thinly spread and British administrative presence in remote regions had been badly affected by the war. In addition, Malayans had seen the British defeated in battle by Japanese forces, but because the war had ended so quickly in 1945 they had not seen the British return to defeat the Japanese (as had happened in Burma). This meant that British prestige in many areas in Malaya was quite seriously damaged in the immediate post-war period. Furthermore, in line with British colonial practice, the lead agency responsible for security was the police, and in Malaya policing was devolved to the states. Thus, command and control arrangements for the key security forces were weak. During 1948–50, a number of changes were made to legislation, command and control arrangements,[2] and to the security forces themselves. But these seemed to have little impact and the insurgency gained momentum.

The Briggs Plan

Born in 1894, Lieutenant General Sir Harold Briggs joined the British Army at nineteen and served with the King's Regiment in France. After a year, he transferred to the Indian Army, serving with distinction in both world wars. In 1946, he was appointed commander in chief, Burma Command, retiring from the army in 1948 at the age of fifty-four. In 1950, Briggs was recalled to active service by Field Marshal Sir William Slim, under whom he had served whilst commanding the Indian 5th Infantry Division in Burma in 1944, and appointed Director of Operations in Malaya—nominally a civilian appointment.

On arrival in Malaya, Briggs quickly reviewed the situation and set to work to produce a plan that is widely credited with providing the conceptual underpinning that enabled the reversal of a serious situation and the eventual defeat of the insurgency. The plan set out to achieve proper administrative control in all of Malaya, starting in the south and moving northwards, and was intended to direct a thorough and long-term operation with no expectation of speedy results. It established joint operational control of the military and police at all levels, close integration of military and police intelligence and set out a minimum level of troops to be achieved across the country in support of the police, who in turn would be responsible for delivering routine security and public safety.

Concentration of force was to be achieved for the clearing of priority areas. The plan also focused on separating the insurgents from their support base and sought to achieve this through the implementation of population control measures. This led to large numbers of Chinese Malayans being forcibly resettled into so-called 'new villages', which were surrounded by ditches and barbed wire, and subjected to movement control orders, extraordinarily restrictive curfews and limitations on access to food, fuel and other essential commodities. These controls were gradually relaxed when areas were deemed to be free of insurgent activity. The plan also introduced a number of initiatives to restrict food supply and medicine to the insurgents and to disrupt the Min Yuen—the MCP's 'Masses Movement', which functioned as the auxiliary support network and provided the connective tissue between the guerrilla main force (operating primarily in the deep jungle and on the edge of Chinese squatter settlements) and the urban militia fronts of the MRLA. The Briggs Plan, as it became known, came into effect on 1 June 1950.

By the end of 1950, sustained pressure by the security forces had compelled the MRLA to split into smaller groups, although the insurgents still appeared to have the upper hand. Interviewed in 1999 at the Australian National University, Canberra, Chin Peng stated that, from his point of view, the high point of the MRLA campaign had been 1948–50. In his memoirs, Chin Peng stated, 'I first heard of Templer's appointment over Radio Malaya. By then we were really feeling the heat of the new villages.'[3]

Towards the end of his time in Malaya, Briggs became increasingly frustrated at his lack of command authority to get the security forces to make the changes he was directing. In late 1951, Briggs retired from the army for a second time, due to ill health. His time in Malaya had taken a toll and he died in 1952, aged fifty-eight. Field Marshal Slim said of him, 'I know of few commanders who made as many immediate and critical decisions on every step of the ladder of promotion, and I know of none who have made so few mistakes'.[4]

The Templer era

In October 1951, the British High Commissioner for Malaya, Sir Henry Gurney, was assassinated in an ambush. This shook British confidence and prompted an urgent review of the situation by the newly elected Conservative government headed by Sir Winston Churchill. The review recognised that the fall of Malaya to communist insurgents would have considerable regional repercussions, concluding that more troops were required, and that the command and control arrangements needed adjustment. Writing to the colonial secretary, Oliver Lyttelton, following Gurney's assassination, the chief of the Imperial General Staff, Field Marshal Montgomery, noted, 'Dear Lyttleton, Malaya, we must have a plan. Secondly, we must have a man. When we have a plan and a man, we shall succeed: not otherwise. Yours sincerely Montgomery (F.M)'.[5]

Sir Gerald Templer was appointed as the British high commissioner for Malaya on 22 January 1952, arriving in Malaya three weeks later. It was a critical moment in the Malayan Emergency, with a large number of insurgent attacks taking place. Templer came to the post with extensive operational experience, having fought in both world wars, and with experience as an administrator, having been in the military government in the British Sector of West Germany after the Second World War. Templer also assumed the appointment of Director of Operations, Malaya, from General Lockhart, who remained as his deputy, thus unifying military and civil command in one appointment. General Templer chose to stick with the Briggs Plan—unifying command and control, improving intelligence, getting the force ratios right, and separating the insurgents from their support base—and drove its implementation with energy and innovation. Importantly, he arrived with instructions from London to prepare the country for independence—something that tackled, head on, one of the main political drivers of the insurgency.

Sir Gerald had a considerable background in military intelligence, culminating in a two-year stint in London as director of military intelligence. He had also served as head of Special Operations Executive's German directorate, giving him experience of clandestine and unconventional warfare, and had experienced insurgency first-hand, being awarded a Distinguished Service Order for service as a company commander during the Arab Revolt in Palestine in 1936. He drove through the centralisation of intelligence, creating the post of director of intelligence to coordinate the actions of all intelligence services. He also directed the establishment of a Special Branch Intelligence Training School, and developed firm links with the intelligence services in neighbouring countries suffering from a communist threat, particularly Thailand. Additionally, he increased the amount of money available to reward people for supplying intelligence, and used collective punishment to get communities to cooperate, most notably at Tanjong Malim. He understood the importance of having well-trained individuals operating with considerable freedom in a counterinsurgency. He encouraged the training and education of military forces through the creation of jungle schools and the publication of doctrine pamphlets. He also drove through a number of reforms to the police force. He travelled extensively throughout the theatre of operations, and demanded results.

The Briggs Plan had called for an increase in Malayan forces and Templer energised this part of the plan. By the end of 1953, eight of the infantry battalions engaged in operations were Malayan; they were well trained and would go on to play an increasing role in the campaign. The police force had grown to 60,000 and had received an injection of experienced personnel following the disbandment of the Palestine Police. The paramilitary Malayan Home Guard was also considerably expanded to a size of 250,000.

Templer handed over his duties as high commissioner to Sir Donald MacGillivray and as director of operations to Lieutenant General Sir Geoffrey Bourne on 30 May 1954. Despite his achievements in reversing the momentum of the insurgency and restoring order to large parts of the country, Templer departed with a warning that the Emergency was still far from over and that tough times lay ahead.

The end of the Emergency

During 1954, operations against the MRLA continued, but by now the guerrilla main force had withdrawn to difficult terrain in deep jungle bases, dispersed into independent platoon structures (akin to FARC's autonomous mobile columns) and were increasingly hard to engage. By the middle of 1954, the monthly total of civilian and military deaths had fallen to thirty-five, in contrast to a monthly high of 188 in 1951. After the July 1955 elections, the new prime minister, Tunku Abdul Rahman, formed a coalition government and engaged in a number of political initiatives designed to end the insurgency. But an amnesty declared after the elections had little effect and talks with Chin Peng broke down in December 1955, with the amnesty ending in February 1956. Military operations continued throughout the year, by the end of which most of the Federation was regarded as safe. This progress continued the following year and July 1957 was the first month in which no people were killed as the result of MRLA action. Independence was granted on 31 August 1957. Increasingly, the insurgents were located across the border in Thailand, with only small numbers operating in Malaya, and the Malayan government pursued a dual strategy of engagement and military action. The most notable success was in April 1958, when 160 insurgents were encouraged to surrender.

The state of emergency was declared over in July 1960. However, the renamed Communist Party of Malaya (CPM) re-launched its armed struggle in 1968. The year 1974 saw an upsurge of insurgent attacks, but also represented a major turning point in the conflict as the CPM split into three rival factions, inflicting considerable internal damage on the movement. Importantly, the then-Chinese Premier, Zhou Enlai, conceded that dealing with the CPM was an internal problem for Malaya, thus signposting that Chinese support would eventually be withdrawn. The conflict dragged on until 1989, the same year as the Berlin Wall fell, at which point hostilities where finally declared over.[6] During this period, the Malaysian government maintained restrictive (and at times controversial) emergency powers under the Internal

16. Map of Afghanistan.

Security Act, which enabled it to act decisively against any guerrilla resurgence. Malaysia's allies also supported the country through the Five Power Defence Arrangements, under which permanent Australian and (until 1989) New Zealand military garrisons were maintained in Malaysia and Singapore to support internal and regional security efforts.

Afghanistan 2001–14[7]

> *Just days after 9/11, Congress authorised the use of force against al Qaeda and those who harboured them—an authorisation that continues to this day. The vote in the Senate was ninety-eight to nothing. The vote in the House was 420 to one. For the first time in its history, the North Atlantic Treaty Organisation invoked Article 5—the commitment that says an attack on one member nation is an attack on all. The United Nations Security Council endorsed the use of all necessary steps to respond to the 9/11 attacks. America, our allies and the world were acting as one to destroy al Qaeda's terrorist network and to protect our common security.*

President Barack Obama, West Point, 1 December 2009

When the US-led coalition launched Operation Enduring Freedom (OEF) in 2001, its aim was the defeat of terrorist elements operating from Afghanistan. It sought to achieve this by defeating Taliban military forces, removing their government from power and destroying those elements of Al Qaeda operating in Afghanistan. The longer-term aim of the coalition was to create a stable, democratic Afghanistan that would no longer be a safe haven for terrorists. OEF started as an unconventional warfare operation using air power in combination with Special Forces and ground troops from the Northern and Eastern Alliance to defeat the Taliban—and this phase was very successful. However, many key Taliban government members and military commanders escaped. An end to combat operations was declared by OEF in May 2003, and at this point the operation focused on stabilising the country. While there had been tactical engagements with small numbers of Taliban in the south and east of the country ever since their fall in 2001, by 2005 the number and size of these engagements had increased considerably. In effect, during the period from 2001 to the start of 2005, the Taliban achieved strategic survival by establishing a safe haven in Pakistan, and then began to run an offensive campaign back into Afghanistan, which they hoped to expand from guerrilla warfare to conventional war fighting.

The International Security Assistance Force (ISAF), as approved in the Bonn Agreement of December 2001, deployed under a UN mandate as soon as the agreement was signed. Initially, ISAF was restricted to Kabul, securing the new government's seat of power. However, the ISAF mandate included scope for assistance in the training of Afghanistan's armed forces, post-conflict reconstruction and, most importantly, for the expansion of ISAF throughout Afghanistan. The ISAF mission was formally taken over by NATO in August 2003, and expansion, initially to the north of the country, started in December 2003 and was completed with the inclusion of the east of the country in October 2006. Until the expansion of the mission to southern Afghanistan in July 2006, the majority of NATO force-contributing nations regarded the operation as a peacekeeping or post-conflict reconstruction mission.

Thus, with OEF seeing Afghanistan as being at the end of a conventional war and ISAF initially viewing Afghanistan as a peacekeeping operation, it was not until the completion of ISAF expansion that the whole international military effort could be put on a common doctrinal footing: that of counter-insurgency. Before 2006, US, British and Australian forces had consciously pursued a COIN approach, as did Canadian forces after mid-2006. But without a common doctrine—which was later enacted by NATO under the rubric of the 'comprehensive approach'—it was difficult for international forces to coordinate action, and correspondingly easy for the Taliban to slip between the cracks to fight another day.

Throughout the period 2007–09, the Taliban-led insurgency grew in intensity, particularly in the south and the east of the country, and ISAF and Afghan commanders made repeated requests for more forces and an expansion in the planned size of the Afghan National Army. By 2008, the Bush administration had committed significant numbers of extra troops to Afghanistan, as the previously dire situation in Iraq began to stabilise and to free up forces for Afghanistan, while throughout 2009 President Obama led a review of the Afghan mission, which resulted in an agreement to surge both US and international forces for an eighteen-month period beginning in January 2010. Beyond this period, there was a clearly stated intent by ISAF nations to disengage from combat operations, the date for which was subsequently set as the end of 2014.

During the jointly UN-, UK- and Afghan-sponsored International Conference on Afghanistan, held in London in January 2010, it was announced that the Afghan National Army would be increased in strength to 171,600 and the police to 134,000 by October 2011, thus providing the scale of forces necessary to enable an international troop drawdown. The hard fighting that followed over the course of 2010 changed the dynamic of the campaign and another US review in December of that year confirmed the start of the drawdown as July 2011, with the end of 2014 set as the date for the end of the NATO mission. The exit strategy was confirmed at the Chicago NATO summit of May 2012 and the NATO combat mission formally ended in December 2014, though US and some allied forces remained under bilateral arrangements. Afghan forces gradually took the lead for security responsibility during the last four years of NATO involvement and, as of 2015, are running the security operation, supported by small numbers of international trainers and a residual US military presence.

The United Nations has been present in Afghanistan since October 1988 with a limited purpose of assistance and coordination, and as a result its footprint has always been relatively small. Following the war in 2001, the UN role in Afghanistan remained one of assistance. In particular, it was felt that the mission needed to operate with a light footprint to prevent it from being seen as yet another occupation, and that an Afghan interim administration should quickly take the lead.

President Hamid Karzai formed the Interim Afghan Administration after the signing of the Bonn Agreement. It was drawn from key members of the Northern Alliance, from Karzai's Eastern Alliance, and from other important groups who had been driven from power by the Taliban. Karzai was elected president in 2004 and re-elected in 2009 for a second and final term. However, very few members of the interim administration and subsequent government had significant experience in running a country, and the civil service, destroyed by over thirty years of war and limited formal government, was ineffective. Furthermore, a democratically elected government had never governed Afghanistan; rather, traditional monarchy or tribal structures had dominated, based on patronage rather than on a culture of service provision to the electorate. The result is that warlordism and illicit patronage networks have remained distinguishing features of Afghan politics. After closely contested elections in 2014 and a subsequent power-sharing agreement, Ashraf Ghani, a former finance minister and chancellor of Kabul University, became president, with his closest rival, former Northern Alliance leader and ex-Foreign Minister Abdullah Abdullah as prime minister and chief executive.

Given the broad international consensus behind the war in Afghanistan, there was much support for the new government from the international community. Lead nations were appointed to help rebuild key elements of the state and support the government in difficult areas. Nations and international organisations offered donations, although many did not want to pass their money through the government and thus funded particular projects directly.

This approach presented the government of Afghanistan with considerable coordination challenges.

A number of themes stand out from the two very different campaigns in Malaya and Afghanistan, and may represent the difference between success and failure in the Colombian government's COIN campaign against FARC. The rest of this chapter examines five of those themes in the context of all three campaigns: the need for 'a man with a plan', security force numbers, hearts and minds and legitimacy, the importance of people, and the length of insurgencies.

The vital ground: a man with a plan

Field Marshal Montgomery's letter to Oliver Lyttelton, quoted earlier, may seem like a statement of the blindingly obvious. Indeed, Lyttelton's response in a note back to Montgomery some days later, 'I may, perhaps with undue conceit say that this had occurred to me',[8] makes this point. But it is a comment that strikes at the heart of what is needed to gain momentum in COIN—whether in an expeditionary context (as in Malaya and Afghanistan) or a domestic setting, as in Colombia. In this respect, Malaya, Colombia and Afghanistan allow for an interesting comparison.

Colombia took a long time to get to the point of having both a 'plan and a man', and bad experiences along the way made the process even longer. In 1959 and 1962, two teams of US COIN specialists visited Colombia to advise the military on how to deal with the remaining areas of conflict after La Violencia; as a result, in 1962, Plan Lazo was launched. The plan contained all the classic elements of a COIN campaign: improving the capability of the armed forces, restructuring intelligence to achieve better coordination, better civil–military cooperation, psychological operations, and operations intended to separate guerrilla groups from their sources of support. Militarily, this plan achieved a number of successes, but it was not applied to the whole of the country, did not achieve true civil–military integration, and strained relations between the military and the government to such an extent that the army commander, General Ruiz Novoa, was forced to resign.[9] Perhaps the key concern for Colombian civilian leaders was that the (doctrinally correct) efforts of military commanders to seek social, political and economic solutions to the basic drivers of insurgency in Colombia's rural periphery looked a lot like an attempt at a military takeover of civil government. Against the background of *coups d'etat* across Latin America in this period, and on the back of a recent period of direct military rule of Colombia under General Gustavo Rojas, this was utterly unacceptable to Colombian political leaders, and the plan was thus overly focused on security to the exclusion of conflict resolution. The plan's critics blamed the resulting heavy-handedness for the formation of FARC in 1964, arguing that it demonstrated the futility of a hard-line approach and the need for a peace process.

With hindsight, it can be fairly argued that the plan had many of the elements needed to defeat an insurgency, and that it was undermined both by being terminated too early and by poor civil–military integration. But the experience coloured the perception of such plans in Colombia, and politicians took a different approach for the next thirty-six years.

As discussed in previous chapters, a workable combination of civil–military measures was eventually reached with Plan Colombia under President Andrés Pastrana, expanding after the election of Álvaro Uribe in 2002. Pastrana began working on Plan Colombia before his inauguration in 1998, the original concept being for a social development plan in support of his proposed peace process. But with the security situation worsening, a request was made to the US government for financial support to the military. The response from the United States was that only counter-narcotics assistance could be provided and that the plan could only be sold to Congress in the context of a comprehensive strategy. As a result, the plan was broadened to include ten elements: economics, peace, development, defence, counter-narcotics, and judicial and human-rights strategies.[10] In this way, the plan came to represent a truly comprehensive approach to dealing with the insurgency and set in train some key reforms to the military, though it was initially orientated towards achieving a peace deal.

Elected with a mandate to sort out Colombia's security problems, President Uribe's style of leadership (as described in the Introduction) was one of energy, personal contact and engagement in the detail—one of micro-management. Uribe's memoir, *No Lost Causes*, offers a feel for his considerable work rate and ethic, and his approach: he would give community leaders his mobile-phone number and tell them to call if they had a problem. As a result, he often knew of problems before his cabinet members and before military leaders. He would then demand results, sacking those who could not deliver, but defending, sometimes controversially from accusations of abuse, those who did. He was uncompromising, but people admired him for his tough approach and he rose in popularity throughout his two terms in office.[11] Having seen the central importance of the need for security as governor of

Antioquia, Uribe saw his mission as president as the creation of 'security with democratic values'. His initial challenge was to give the people of Colombia the confidence that this could be achieved. He sought quick wins to show the people what could be done.[12] The rapidly implemented policy of allowing people to travel between cities on public holidays in military guarded convoys known as 'caravans' was extremely successful in this respect, and from these small gains he built momentum.

Colombian governments after Pastrana achieved a striking degree of continuity across administrations. With regard to Plan Colombia, Uribe, for his part, did not go back to the drawing board as new administrations so often do; rather, he built on what had come before. After Uribe, President Juan Manuel Santos, who had served as one of Uribe's defence ministers, kept momentum on the plan going. When progress appeared to falter in 2010 through lack of focus, the military campaign plan was extensively reviewed under the direction of newly appointed Minister of Defence Juan Carlos Pinzón and the resulting Sword of Honour campaign plan, which concentrated on dealing with those areas that had not been properly delivered, re-established the momentum, as described in Chapters 1 and 3.

In Malaya, the process had been much quicker and—since Malaya was, after all, a British colonial possession—democracy was not a factor in the appointment of the 'man'. The situation had deteriorated rapidly since February 1948, prompting a series of actions, including the appointment of Briggs as director of operations. Gurney had not been idle, introducing a series of important reforms and pieces of legislation in 1949, and the Briggs Plan was written in the early part of 1950. The assassination of Gurney combined with an incoming British Conservative government, which was faced with an array of serious challenges—not least the potential for the spread of communism in the Far East and the war in Korea. This meant that the situation in Malaya too was given serious attention in London. Both Field Marshall Slim and General Robinson were approached about taking on the appointment of high commissioner. Both declined primarily due to age. Their consideration showed how seriously the government took the appointment and reveals the quality of people they sought to take the post.[13]

Once selected, Templer was asked by Colonial Secretary Lyttelton what he needed. He replied that he wanted a clear directive from the government giving him a mandate, and once in Malaya he set about his role with considerable energy. He too had a business-like, abrupt, almost rude style that demanded results, and those that did not learn quickly were dismissed. Like

Uribe in Colombia, he toured the country constantly to see things for himself. He was ruthlessly efficient, but balanced this with a willingness to apologise for his own mistakes.[14]

In Afghanistan, the challenge was of a different order. There was no functioning government whatsoever after the fall of the Taliban, and the country was in a state of considerable disrepair. Indeed, it can be argued that Al Qaeda and the Taliban understood the potential impact of having a strong leader in charge of a post-Taliban Afghanistan: AQ-linked terrorists assassinated Ahmad Shah Massoud, leader of the Northern Alliance, two days before 9/11 and the Taliban also managed to assassinate the widely respected potential future president Abdul Haq on 26 October 2001, a man regarded as one of the few leaders capable of restoring unity to the country. The new interim administration established under the Bonn Agreement was a compromise designed to balance representation and power. At the time of the administration's appointment and endorsement by a Loya Jirga 'or grand tribal council' in 2002, the Taliban was not regarded as a major threat to national security; the challenge was one of reconstruction and, indeed, the Quetta Shura—later the key leadership group of the Afghan Taliban—was not formed until mid-October 2003. By the time of President Karzai's election in 2004, although security was becoming a major issue, the challenge was still conceived primarily as one of reconstruction. Indeed, in the early years of the war, the various military forces engaged in Afghanistan, as already noted, lacked a common approach and viewed the campaign at different points along a spectrum from war to peacekeeping.

Even if Karzai had been the 'man', however, he had a bewildering array of plans to choose from, all designed by outsiders, all highly complex and none owned or driven by the Afghan government. Following his 2009 re-election, when the violence was reaching new heights and the US and NATO were gearing up for a surge of the security effort, relations with the US reached a low point. Karzai's consistent approach to the conflict was to push for peace, to attempt to negotiate with the Taliban, to pardon combatants, to blame Pakistan (with some justification) for the Taliban's operations from its territory, and to impose limits on the tactics of international forces so as to reduce their negative impact on the civilian population. At best, he seemed to regard the security effort as a necessary evil until another way could be found to end the conflict. As he neared the end of his second term in office he became increasingly critical of international forces, arguing that they had done more harm than good.

Perhaps the best attempt to create harmony between 'a plan' and 'the man' in Afghanistan was the Policy Action Group, encouraged by General David Richards, ISAF commander, in 2006. This initiative tried to put a war cabinet, chaired by Karzai, at the helm of the campaign, thus unifying all elements of international support behind the government. Ultimately, it failed because it was not fully embraced by either the Afghan government or the international community. At the heart of the problem was a lack of trust between the two, the touchstone issue being that of corruption. Accusations and counter-claims abounded. As a result, for example, many large international donors would not allow their money to flow through the Afghan government and thus local communities did not see the resulting projects as instances of the new government delivering services to its people.

The comparison between Malaya and Colombia on the one hand, and Afghanistan on the other, is stark. As a colony of the British Empire, in Malaya finding the 'man' could be done quickly and the need for a plan was obvious and undisputed, as Lyttelton had observed. In Colombia, the peace process engaged in by Pastrana showed the country that FARC was not serious about peace, thus setting the stage for Uribe to be elected with a mandate to be tough on security. At the same time, Pastrana had laid good plans with Plan Colombia, on which Uribe was able to build. In Afghanistan, meanwhile, neither Karzai nor current President Ghani was elected for his policies, and Afghan democracy remains too immature for such a national debate. Tribal and ethnic identities, and associated vested interests, are too deeply ingrained. The selection of General Abdul Rashid Dostum, a notorious ethnic Uzbek Northern Alliance leader, as Ghani's vice president, illustrates the point. In 2009, Ghani had lashed out at Karzai for bringing Dostum back into politics, describing him as 'a known killer'.[15] Explaining his subsequent choice of Dostum as a running partner in his own presidential bid to the press, Ghani said: 'Politics is not a love marriage, politics is a product of historic necessities'.[16]

There is a (quite well-founded) feeling in Afghanistan that the war since 2001 has been a war for the benefit of outsiders who are not really concerned for the people of Afghanistan, and as a result the conflict has lacked the strong local ownership and leadership that represents the vital ground in winning such a conflict. Meanwhile, Afghans themselves experience a competing set of pressures and priorities, many of which are ethnically or historically driven. As one illustration of this, an Afghan minister, who was an ethnic Pashtun, speaking in 2007 about the country's border with Pakistan, agreed that official recognition of the border would take a lot of tension out of the Afghan relationship with Pakistan and help to counter the insurgency, but explained that such recognition would, however, amount to the end of the dream of a greater Pashtunistan—an area of land spanning the border and an idea that threatens the integrity of Pakistan. No Pashtun politician could agree to such a deal and survive politically; hence the issue rumbles on unresolved, as does the war. The contrast been Malaya and Colombia both of which arrived at a point at which a man, with a mandate, and a sound plan led the campaign and Afghanistan where this has yet to truly happen is stark.

Quantity has a quality all of its own

Numbers of 'boots on the ground' became a hot topic in the recent war in Afghanistan, and a quick scan of the Internet reveals just how much has been written on the topic. It was also a major feature of the COIN campaigns in Colombia and Malaya.

It became an issue in Afghanistan because international troops are costly, both in terms of cash and political capital, while local troops take time to generate if they are to be trained to the right standard. However, without the right numbers, it is impossible to secure a population from insurgents. Each campaign will be different, for the number of troops required will depend on the size of the population and its geographic dispersion, and the physical size of the country and the nature of the terrain. It is not as simple as establishing a ratio of population size to the number of security forces required. Comparisons of the three campaigns illustrate this issue clearly.

In 1947, the population of Malaya was assessed at 5 million, of whom 2 million were ethnic Chinese, half a million ethnic Indians and the remainder ethnic Malays. The Malayan Peninsula covers 131,598 square kilometres, although at the time of the Malayan Emergency, 90 per cent of the population lived on the coastal plain extending 10 miles inland along the western coast of the peninsula. At the start of the insurgency in 1948, there were about 10,000 police officers and twelve infantry battalions (comprising 9,600 men) stationed in Malaya. Security force numbers were ramped up quickly and reached a peak in 1952 at 60,000 police, 40,000 army and 250,000 Home Guards—a total force of some 350,000. At its peak, the MRLA had an estimated 7,000 fighters and 30–40,000 active supporters. The scale and pace of the expansion of the security forces to counter this force was impressively fast.

By contrast, a 2013 estimate put the population of Afghanistan at 28 million, of whom approximately 45 per cent are thought to be Pashtun. Afghanistan covers an area of 652,225 square kilometres, but big tracts of the country are uninhabited desert or mountainous areas, with large elements of the population concentrated around the major rivers. In 2011, at the height of the international troop surge, there were 132,000 international troops, 152,000 members of the Afghan National Army and 118,000 members of the Afghan National Police deployed—a total of 402,000 personnel.

In Colombia, meanwhile, the population was estimated at 39.9 million in 2000, while today it is closer to 47 million. As described in previous chapters, Colombia is vast, covering an area of 1,138,903 square kilometres, while topographically the country is split into four regions: the central highlands, the Caribbean lowlands, the Pacific lowlands and eastern Colombia, with roughly 94.5 per cent of the population concentrated in just 42 per cent of the country. The eastern plains are sparsely populated, while 75 per cent of the population live in the central highlands. Against this enormous area and uneven population distribution, in 1998 the strength of the army was 104,000—a mix of conscripts (80 per cent) and long-service professionals—while the national police stood at 90,000. By 2014, these numbers had grown to 237,000 for the army and 181,000 for the police, alongside 33,000 for the navy and 15,000 for the air force—a total strength of 466,000.[17]

Whilst it is hard to draw any direct comparisons between these three given technological advancements and the considerable geographical and population differences, in each case the security forces had to be expanded considerably, reaching the 350–450,000 mark in all three. Rapidly raised local forces provided the bulk of the expansion in each case. This analysis raises three issues: COIN norms in relation to force numbers, the quality of rapidly raised local forces, and the staying power of international forces—which will be addressed in the remainder of this section.

In general, at the start of a campaign, it is very difficult to calculate the numbers of troops and police required to defeat an insurgency, because the forces engaged are in a rapid learning period, making it hard to develop robust arguments for the force size required. In effect COIN norms in relation to force numbers have yet to be established. This inability to make a strong argument for force numbers is set against a political desire to keep the numbers as low as possible to do the job. The result is generally incremental growth that lags behind the growth of the insurgency, and a consequent lengthening of the campaign. Colombia and Afghanistan both demonstrate this tendency. In Afghanistan, debates over troop numbers and the size of the Afghan National Army and Police became highly toxic and both ISAF Commanders Generals

David McKiernan and Stanley McChrystal did not get the US troop numbers they wanted for the surge. With regard to local forces, meanwhile, as early as 2006, the Afghan defence minister, General Abdul Wardak, was arguing that the plan to create an Afghan army of around 70,000 (a number based on what it was estimated the country could afford in the long term) was inadequate, and that the army was in need of considerable expansion to allow the combat load to be carried by Afghan and not international forces. Such expansion plans were not agreed, however, until early 2010 at the London Conference.

Looking at this issue through a tactical lens is illuminating. During the planning for Operation Moshtarak (an operation designed to restore security to Helmand Province), an extensive study was undertaken[18] to find out what troop numbers were really needed in the terrain of the Helmand Valley. It was determined that an ISAF company (120 men) partnered with Afghan National Army and Police (200 men) could secure an area of 10–12 square kilometres. The surge had provided the British force with twenty-five companies. This meant that UK forces could hold about 275 square kilometres, the size of central Helmand only. On this basis, the North Helmand Valley and areas of southern Helmand were handed over to the US Marines and UK forces concentrated in central Helmand, where the vast majority of the population lived. Once this had occurred, company commanders had the resources to apply COIN tactics and techniques properly, and security in central Helmand improved dramatically over the following 18 months. The problem was that this planning yardstick took time to develop. It also meant that low-priority areas had to be given up in order to achieve proper effect in populated areas, something that carried a political cost. Also, as noted in Chapter 3, a counter-guerrilla approach that sees periodic visits by military forces to a given area can expose, and then destroy, those members of local populations who support the government—and this effect is even more pronounced if an area that has previously been occupied by the security forces has later to be abandoned.

For quickly raised local forces, meanwhile, the biggest challenge relates to their rapid training, and their resultant quality and political loyalty. In its early days, the Malayan Home Guard—in effect, a government-sponsored paramilitary self-defence militia—contained many communist sympathisers and, as a result, substantial quantities of arms and ammunition were channelled to the insurgents. In response, they were subjected to a rigorous screening process, which gradually improved the situation. Likewise, in Afghanistan numerous local police and home-defence initiatives were tried, but many were quickly tainted by corruption, allegiance to local powerbrokers and a lack of respect for human rights. These were failures, in the case of Afghanistan, that the international media was quick to point out and the Taliban eager to exploit.

In Colombia, the growth of security forces was slower, training more rigorous and, as a result, the problems within the formal security forces fewer. At the same time, however, the emergence of the *autodefensas* (paramilitaries), as populations expressed their lack of confidence in the ability of the Colombian state to protect them and instead took up arms to defend themselves, created massive violence and disruption that was only partially resolved when the *autodefensas* demobilised in 2006, and still persists today in the guise of BACRIM.

All three campaigns differ markedly in the role played by international forces—and in relation to their staying power.

Malaya was part of the British Commonwealth during the 1950s (and, indeed, remained so after the end of the conflict) and Commonwealth forces were rapidly deployed to bolster security, remaining in greatly reduced numbers even after independence had been declared—indeed, Australian troops and warplanes are still stationed in northern Malaysia today. Yet this setting was unique, the conflict taking place shortly after the end of the Second World War, and concurrently with the Korean War and the rise of communism in the Far East. In Colombia, by contrast, international assistance has been on a far smaller scale, with a lower profile, much of it advisory and equipment-related, and linked to the more politically neutral topic of counter-narcotics. As a result, however, it has been long-running, with the first US military survey team visiting the country in 1959 and continual engagement ever since. In Afghanistan as the campaign intensified and lengthened, casualties increased and governments changed, domestic support for the deployment of western troops weakened, leading to calls for a timetable for withdrawal and a deadline for engagement in combat operations, which was set as the end of 2014.

What expeditionary, third-party campaigns such as Afghanistan and Malaya highlight is that when large numbers of international troops are deployed, they are time-limited, susceptible to changes in their domestic political scene, and are, at best, a bridge to a local solution. In this respect, the way in which the Colombians—operating in their own territory, against a domestic adversary, and in control of their own fate—have generated the force numbers required for the campaign and integrated the international support offered bears study, for it shows what can be done with well-targeted assistance under strong local ownership.

From 'hearts and minds' to a battle for legitimacy

If there is one phrase that General Templer is remembered for, it is his call for a campaign that focused on winning the 'hearts and minds' of the Malayan people. If he could win over Malayans, he believed, support for the MCP would dwindle and the insurgency would slowly die out. Winning 'hearts and minds' is about making an intellectual and/or emotional appeal to win people over to your cause. The theory is simple, the practice much more difficult.

This is because such an approach requires every person working to defeat the insurgency to consider their actions in this context. Failures are exposed and publicised, increasingly so given modern communications and the reach of the media. Indeed, there were several incidents that took place during the course of the Malayan Emergency that today would have had (negative) strategic consequences for the campaign, such as the Batang Kali shootings in December 1948, when twenty-four Chinese Malayan rubber plantation workers were shot and their village burnt—a case that is still making news headlines in 2015.[19] In the Malayan Emergency, Templer's focus was very much on the domestic audience. But today, 'hearts and minds' has become a much wider concept, which includes a significant international dimension, as the campaigns in Colombia and Afghanistan illustrate with their need for international support and coverage in the media.

At its core, in a COIN context, winning 'hearts and minds' is about the legitimacy of the national government. If the government is widely seen—both domestically and internationally—to have been fairly elected and to govern in accordance with domestic and international law in an even-handed manner, there will be little support for an insurgency domestically or internationally. The German political philosopher Dolf Sternberger defined legitimacy as 'the foundation of such governmental power as is exercised, both with a consciousness on the government's part that it has the right to govern, and with some recognition by the governed of that right'.[20]

In Colombia, FARC argues that the right system of government for the country is a socialist one, and has campaigned for this through violent revolution. Concurrently, the international community has been highly critical of human rights abuses allegedly committed by government forces. This became a strategic issue during the war, for not only did such alleged abuses alienate the civilian population and undermine the legitimacy of the government, they also weakened international support for the fight against FARC and other guerrilla groups. Indeed, in 1996 and again in 1997 the US government decertified the Colombian government for aid on the basis of its human rights abuses, as well as the alleged corrupt practices of then-President Ernesto Samper Pizano who, as mentioned in Chapter 4, was accused of accepting campaign donations from drug traffickers.

Such lobbying continues today. Take, for example the following statement from Amnesty International: 'Amnesty International USA has been calling for a complete cut off of US military aid to Colombia for over a decade due to the continued collaboration between the Colombian Armed Forces and their paramilitary allies as well as the failure of the Colombian government to improve human rights conditions'.[21]

Successive Colombian governments have recognised this issue and the potential for the loss of both domestic and international support, and have tried to address the situation. In 1961, Colombia ratified the Geneva Conventions of 1949 that form the basis of international humanitarian law. The additional protocols of 1977 were ratified in 1993 and 1995, respectively. The 1991 Colombian constitution details the fundamental rights of citizens, which are protected under national constitutional law. The 1991 constitution also established a central government human rights agency.

Indeed, it was President Samper who changed tack from previous administrations and admitted that Colombia had a problem. He established the Trujillo Commission to investigate the killing of sixty civilians by paramilitaries and accepted its findings, although the commission failed to bring about legal accountability. Looking back, Samper saw this as the start of 'a consciousness-raising campaign with the military', whereby 'human rights offices were opened in all the departments, a special human rights advisor was named, and a military human rights curriculum was established'.[22]

The 'false positives' scandal, a series of attempts to inflate the statistics for guerrillas killed by targeting civilians and calling them guerrillas to meet targets, was a strategic shock for the government and drove further change. The scandal came to light in 2008, when twenty-two men from Soacha, who had been recruited for 'work', were found dead. The recruiter testified, in court, that he had been paid money for each man by the military. In August 2009, UN Special Rapporteur Philip Alston reported:

> I have found no evidence to suggest that these killings were carried out as a matter of official Government policy, or that they were directed by, or carried out with the knowledge of, the President or successive Defence Ministers. On the other hand, the explanation favoured by many in Government—that the killings were carried out on a small scale by a few bad apples—is equally unsustainable.[23]

As at June 2012, a total of 3,350 such cases had been investigated and verdicts reached in only 170 cases.[24] These failures have been capitalised upon by

FARC, despite its own serious abuses, in its efforts to gain international recognition and support. Acting in accordance with the law always be a challenging issue for soldiers and policemen whose own families and colleagues have suffered at the hands of the guerrillas for they are being asked to put aside their personal feelings; some will inevitably fail this test. Correct and repeated training and education in this area is essential as one of the keys to success lies in the security forces maintaining the moral high ground, even when confronted by a ruthless and immoral adversary.

The 2013, the UN General Assembly's Annual Report on Human Rights in Colombia highlighted 'positive developments fostered by the Government of Colombia to promote and protect human rights in 2012', but noted that there was still much to be done.[25] The Human Rights Watch report for 2014 on Colombia noted that 'There has been a dramatic reduction in cases of alleged unlawful killings attributed to security forces since 2009; nevertheless, some isolated cases were reported in 2012 and 2013'.[26] Continued progress will be required for Colombia to change its image and attract the investment and tourism it needs to boost its economy.

In Afghanistan, meanwhile, legitimacy, as should be expected, was an issue of huge significance and created considerable tensions at the heart of the campaign. Given the make-up of the coalition, the size of the international troop contribution and, for some nations, the unpopularity of the war in Iraq, maintaining support for the war at home was challenging. Initially, support for the war in the US and Europe was very strong, as the above quote from Obama's West Point speech highlights, and Afghan support for the removal of the Taliban was also high. But narratives directed at Afghan and international audiences, at times, worked against each other and when things became difficult it was the message to the international audience that dominated.

Consider General McChrystal's directive to use 'courageous restraint': the need to take risks, even to the point of physical harm, in order to prevent damage to the mission. In this directive, he was saying that ISAF needed to reset its rules of engagement; that it was better to exercise caution, if there was any doubt, and for example let a possible insurgent get away rather that kill an innocent civilian and turn a community against ISAF. In the context of winning 'hearts and minds' in Afghanistan, this seems perfectly reasonable. Indeed, Karzai had been constantly criticising ISAF for causing civilian casualties, demanding changes to the way in which air strikes and night raids were conducted. Yet googling the term 'courageous restraint' throws up a host of international media articles claiming that ISAF's soldiers' lives were being endangered by such a policy—articles that further undermined the coalition's domestic support.

Note also the narrative around the drawdown of international troops: one designed to reassure a weary international audience that this was not a war without end. The effect on the Afghan population, meanwhile, was extremely unsettling, with many beginning to hedge their bets as how to prepare for the international withdrawal, and reconsidering their support for the Afghan government. This was not a message calculated to win the 'hearts and minds' of Afghans. An Afghan government on its own was a considerably different proposition to that of an Afghan government with very strong international support.

Consider, too, the reaction of the Taliban, who were effectively being told to hang on for a couple of years and helpfully being told exactly when international forces would be gone. We now know that this issue was a major factor in the Taliban's decision to delay the announcement of the death of their leader, Mullah Omar. By contrast, when Templer took over as British high commissioner in Malaya, his statement setting out the intentions of the British government asserted that Britain would not lay aside her responsibilities in Malaya until the government was satisfied that communist terrorism had been defeated; there was to be no repeat of the situation in Palestine. Afghanistan and Malaya thus offer quite a divergence in approach.

Yet today, as noted, any COIN campaign has to consider not only the 'hearts and minds' of those directly touched by the insurgency but also those of the international community, particularly those countries with influence. If the government starts to lose legitimacy, domestically and/or internationally, as a result of the conduct of a COIN campaign, the consequences need to be understood and mitigating action quickly taken. Mixed messaging needs to be avoided. Then president elect Uribe's report of his discussion about the capture by FARC of former Colombian senator, presidential candidate and anti-corruption activist Ingrid Betancourt (who also held French citizenship) with then-President Jacques Chirac in July 2002 captures this issue perfectly. "'I understand", Chirac replied. "But perhaps you could just give them a few words upon your departure [from France] to hearten [the French public]." In reference to which Uribe reported: 'I shook my head. "I have to say the same things here as I say in Bogotá", I said. "I cannot change my speeches based on the audience. I have to be consistent in private and public."'[27] Modern communications have amplified this aspect of COIN considerably.

The importance of people

One of the greatest contrasts between the campaigns in Colombia and Afghanistan, however, lies in the development of human capital. In Afghanistan during the years of fighting and chaos, those with an education left the country in large numbers; the education system collapsed and government all but ceased to function. A major achievement since 2001 has been the rebuilding of the system and returning large numbers of children, both boys and girls, to school. For many of those in their twenties and above, the damage has been done, yet these are the Afghans on the frontline of dealing with the insurgency. War amongst civilians is a complex, subtle undertaking that requires both soldiers and officers to have an understanding of the theory and practice of COIN warfare, and a real feel for their commander's intent. But many in the rank and file of the police and the army cannot read and write, let alone speak both Dari and Pashto. In 2006, an Afghan government minister who was frustrated with the pace of progress said, 'I have 5,000 civil servants, but I only have twenty who can do any real work'. The international community has stepped in to help and people have come back from abroad to serve their country, but neither source can provide in full the volume of human talent required. For Afghanistan, this part of the recovery process will be measured in generations.

In Colombia, things could not be more different. Sitting around a table with fifteen police officers in Medellín in early 2015, the authors were pleasantly surprised to discover that every single one had a first degree and many had a master's-level qualification. This, they said, was due to an initiative the police force started in 2004 to develop its employees. Army officers receive a similar investment in their education and development, with many being sent abroad to study alongside other nations' armed forces, particularly those of the US. Indeed, the Colombian War College dates back to May 1909, having been formed as part of a move by the then President Rafael Reyes to professionalise the officer corps. The challenge of having large elements of the army provided through conscription is dealt with through regular continuation training across a conscript's whole period of service. The investment in the development of those fighting the insurgency has been huge, and it shows.

In Malaya, the federal government benefited from a British civil service distributed across the territories, the presence of a British representative alongside each state leader, and a network of British officers in the army and police force. Thus the high commissioner and director of operations at least had a framework of human capital spread across the country. Considerable effort was put into finding the right people to lead Malaya through the Emergency and for this the British government had a wealth of talent and experience at its disposal with which to underwrite the campaign. For example, as part of his review in the late autumn of 1951, Lyttelton asked for the resignation of Colonel William Gray, the head of the Malayan Police, and replaced him with Colonel Arthur Young, the commissioner of the City of London Police. Young was perhaps the most distinguished British colonial police officer of the twentieth century, with extensive experience in colonial policing in Africa, later service in COIN environments in Kenya and Northern Ireland, and experience in post-war policing during the military occupation of Italy (where he reformed Italy's Carabinieri and created the modern constabulary police concept). Young's appointment gave Templer an essential and like-minded partner. Templer also brought in Jack Morton, the second in command of MI5 (the UK's domestic intelligence organisation), to help him reorganise Malayan intelligence.

The key deduction must be that the development and correct employment of human capability is vital to winning a counterinsurgency campaign. It follows that the bulk of the talent will have to be both found and developed in-country, and that insurgencies are generally long enough for good plans, laid early on, to address critical shortfalls and affect the outcome of a campaign. Finally, however, while it is clear that international support will rarely generate large numbers of talented people to support a campaign, small numbers of well-placed individuals, particularly in specialist areas, have nevertheless been shown to have a considerable effect.

Insurgencies rarely end quickly—they tail off

Perhaps the most instructive lesson from Malaya is how long it took to really bring an end to the insurgency. Against a background of considerable communist activity in the Far East and the war in Vietnam, between 1968 and 1973 the CPM infiltrated back into the Malayan Peninsula and slowly re-established its support base. 1974 saw an upsurge of CPM attacks. However, the 1969 Sino–Soviet split and the subsequent diplomatic overtures made by the Chinese towards non-communist countries in the Far East had major implications for the CPM, with financial and material support eventually cut off in 1980. The second Malaysian emergency developed into a low-level conflict of subversion and small jungle skirmishes. But by 1988, the war had finally gone against the CPM for good. With the increasing prosperity of both Singapore and Malaysia,

---

the CPM's claims to be liberators from oppressive neo-colonialist regimes no longer struck a chord with the public. At the same time, the guerrillas' safe haven in southern Thailand was gradually cut off as a result of the Royal Thai Army's increasing success against its own communist insurgency. The violence used by the CPM only served to further alienate the public. Chin Peng agreed to a peace treaty and a formal end to the conflict in 1989.

The main lesson from the Malayan Emergency is that the insurgency only really ended once the rapid economic progress made by both Singapore and Malaysia in the 1970s and 1980s had given the vast majority of the population of both countries a new stake in their futures, and external support for the CPM had dropped away. Indeed, the immediate trigger for the final surrender of Chin Peng and other former guerrillas in 1989 had very little to do with events in Malaya, and more to do with the fall of the Berlin Wall and the collapse of international communism. With regard to both Afghanistan and Colombia, this suggests just how much still needs to be done.

In Afghanistan, the Taliban have been knocked back in their ambitions to return to power, but are still far from being defeated militarily and remain a force to be taken seriously. Both the Afghan National Army and Police have grown quickly to fill the gap left by the departure of international forces, but an extended period of support and development is needed to strengthen these institutions to a point where they are self-sustaining and self-correcting. The politics of the country are still fragile, with voting conducted largely along ethnic lines and the president having to strike numerous compromises to exercise his power.

Afghanistan's economy has recovered significantly since the fall of the Taliban regime mainly due to considerable international aid, reform in the agricultural sector and growth in the service sector, but growth slowed in 2013 and may have slowed again in 2014. The country remains poor and very underdeveloped. Not surprisingly, the government in Kabul faces some huge problems, including poor national infrastructure, high unemployment and low levels of job creation, corruption, a sizeable drugs trade and linked black economy, weak government capacity and low rates of tax collection. In Kabul and the big cities, there is visible economic development, and the traffic jams to match. In the rural areas, which remain deeply conservative in outlook and have seen much of the fighting, less has changed. Indeed, many of the quick impact changes have been delivered and what is now needed is sustained employment growth in the formal economy. Yet delivering this will be extremely challenging and require long-term focus and effort.

Furthermore, despite the recent growth of the Islamic State's so-called Khorasan province stealing the radical limelight, the Taliban still has strong international backing. They remain able to operate from Pakistan and enjoy considerable financial and logistical support from a range of wealthy Middle Eastern backers. Until this support is curtailed, they will remain a serious threat to the stability of Afghanistan. And, despite a number of overtures from the Afghan government, the Taliban have not yet engaged in serious peace talks and remain internally divided on the benefits of so doing.

Yet such talks are essential if a return to peace is to be achieved. Perhaps the biggest mistake of the entire campaign was not allowing the Taliban to be represented at the Bonn Conference in 2001, and hence excluding it from the new, post-9/11 Afghanistan. Taken together, all of this suggests that the end of the war in Afghanistan is some way off. To survive, the Afghan government will need serious international support over the long term and a deal with Pakistan. Without it, there will be a slow descent back into chaos.

In Colombia, the situation is different, but no less complex. FARC is very close to being militarily defeated. It has been driven away from the population towards the margins of the country and, as a result, the war against FARC now touches the lives of only very few Colombians. Its leadership has been actively targeted, and international support for its activities weakened. As we explained in detail in Chapter 5, for FARC it makes complete sense to take the struggle back into the political space while it still can. After all, the M-19 terrorist group laid down its arms in the late 1980s, then received pardons from the government and became a political party. Gustavo Petro, a former member of the group, became a senator in 2006 and mayor of Bogotá in 2012.

The challenges and risks for the Colombian government are huge. There will be strong pressure to deliver peace before 2018, the end of Santos's second and final term in office, and the dividends shortly thereafter. Once it has been negotiated it will have to be paid for and delivered on the ground. And, for the peace to last, economic and political progress will be required to remove the underlying drivers of the insurgency. Here too the tail of the insurgency is likely to be a long one.

CONCLUSION

PROSPECTS FOR PEACE AND WIDER IMPLICATIONS

Dickie Davis, David Kilcullen and *Greg Mills*

The port city of Buenaventura, on Colombia's Pacific coast, is a microcosm of the country's social, economic, and security challenges; challenges that have a resonance with many developing countries, particularly those in Africa.

The port is fast expanding, with international investment helping meet the demands of Colombia's economy, which has been persistently growing at around 5 per cent annually for the last decade. But in other respects, this is a town that time—and the government—forgot. One senior naval officer says criminality and unemployment are the result of 'decades of neglect by the central government', going back centuries, when Africans enslaved by Spanish colonists escaped and settled the Pacific coast—a region that is still more than 90 per cent Afro-Colombian.[1]

Buenaventura's 400,000 inhabitants have long been caught in the crossfire of a murderous gang war between the Urabeños—a nationwide criminal group—and its rivals, La Empresa. The two gangs compete for control of this critical drug-trafficking node, which, via a two-hour truck ride, connects the Pacific (and producers further south) with Colombia's most dangerous city, Cali. Hundreds have been killed, and thousands displaced by the violence of the last fifteen years.

Community leaders have been victims of *descuartizando* (murder by quartering), a technique of extreme intimidation: eleven dismembered bodies were

179

found in the first five months of 2014 alone. The *barrio* of Piedras Cantan, its houses built on stilts over the harbour like a miniature version of Makoko in Lagos, has seen as many as twenty-seven murders a month for a population of only 220 families. By 2014, 500 troops of the Colombian Navy's Marine Infantry were working in support of 1,200 police to rid the city of crime and violence—part of a wider struggle against insurgents and lawlessness that has ebbed and flowed over fifty years.

In the process, Colombia has had to learn how to extend governance, security, and infrastructure across a large, rugged territory, one-third of which is covered by jungle so dense it looks like broccoli from above. The absence of governance and essential services has intersected historically with a lack of licit job opportunities, encouraging criminality from narco-trafficking to illegal mining.

Yet things have changed dramatically, and for the better. For decades, Colombia was metaphorically cut off from the outside world—the land of Pablo Escobar, its lifestyle defined by the term *blanco y negro* ('white and black', a line of coke and an espresso), and its political economy by the Medellín drug boss's question, *plata o plomo?* ('silver or lead'?) Back then, just fifteen years ago, tourists and business travellers steered clear of Colombia because of its image of drug cartels, guerrillas and kidnappers. Now they are back in rapidly increasing numbers.

At the turn of the century, with the failure of a peace plan in 2000, just a third of Colombia's territory was completely under government control: FARC forces threatened most Colombian roads, operated in groups of up between 200 to 1,000 guerrillas, and ran a narco counter-state. Now the government controls over 90 per cent of its terrain, violence and kidnapping are down dramatically, and only eighty municipalities (of roughly 1,100 nationwide) see significant FARC activity, suggesting that the threat to Colombia's people has been significantly rolled back.

Colombia is confronting its issues of rudimentary service-delivery, while simultaneously dealing with armed challenges to state authority. Much still remains to be done. But for those facing similar challenges—as in eastern Congo or Mali, or for those facing Boko Haram in Nigeria, Chad and Cameroon, or Al Shabaab across East Africa—there are valuable lessons to be learnt from Colombia.

Processes for consolidation

As this book has argued, improved security—through effective military counter-guerrilla efforts complemented by improvements in human rights, legitimacy, and essential services—has been the door through which all else has followed, including growth and development fuelled by an increase in investment and confidence. FARC and other guerrilla movements that once beat at the gates of Bogotá have been pushed back into the jungle and regional hideouts by an emboldened and strengthened military, itself a product of a motivated political leadership. This has transformed Colombia from what its then defence minister Juan Carlos Pinzón has described as 'almost a failed state' into a more stable, more prosperous, and much more secure society. At the same time, however, as we note, unless civilian agencies bring effective programmes to complement the work of the security forces, and unless stabilisation measures are sustained for a very significant period of time, the risk is that, after all this blood and effort, Colombia will simply have re-established its historical dynamic of stalemate in which guerrillas cannot overthrow the central state, and yet the state cannot eradicate persistent, illegal armed groups from the country's remote periphery. Thus, whether the turnaround achieved since 2002 can be sustained depends on keeping markets open, ensuring peace, institutionalising progress, and deepening Colombia's democracy.

Despite the popular image of omnipotent guerrillas and drug lords, Colombia boasts the longest-running uninterrupted democracy in Latin America—fifty-seven years, versus twenty-four years in Chile and twenty-nine years in Brazil. In comparison with a region that has suffered more than fifty *coups d'état* since 1940, Colombia experienced just one four-year period in the 1950s under military rule.[2]

The Santos administration's approach to security and development has meant the onus no longer lies exclusively on the armed forces, but rather on an alignment of citizen safety and security with development and economic modernisation. Institutions are stronger and are deepening their impact across the broad expanse of Colombia's territory. This is especially relevant in the many remote rural departments where security and public services have been lacking—such gaps in institutions and public services have proved fertile ground for exploitation by the guerrillas. A broadly functional Colombian state and its institutions are now more actively involved and visible to all citizenry, leaving little reason for opposing—and often violent—forces to creep in.

Whereas Colombia was number one in the world in terms of kidnappings and crime rate at the start of the twenty-first century, by 2014 these measures fell by 95 per cent and 43 per cent respectively. Such dramatic improvements have been fundamental to the turnaround in business and tourism. Apart from encouraging the movement of local Colombians and foreign visitors around the country—armed escorts used to usher holidaymakers from one part of the country to another—a safer environment has brought with it foreign capital, more opportunities, and rapid development. Research suggests that the movement of people and information is a key driver of sustained economic integration and growing connectedness.[3] This is clearly the case in Colombia where Internet users have grown from 2.2 per cent of the population in 2004 to 52.6 per cent ten years later.[4] Similarly, the percentage of Colombians with mobile phone subscriptions has risen from 24 per cent in 2004 to 113 per cent over the same timeframe.[5]

Apart from the security imperative complemented with the right mix of policy and institutional reforms, Colombians themselves have taken hold of their national destiny. Long known for their resilience and warmth, Colombians are the second-happiest people on the planet according to the global Gallup happiness poll.[6] A positive attitude combined with a growing number of Colombians returning from abroad, and improving skills and education, mean human capital in Colombia is in good shape and bodes well for the future.

El Dorado International Airport, the new airport in Colombia's capital, bustles with visitors from across the globe. Built around the old terminal—a shell still remains as a reminder of Colombia's less prosperous and difficult past—this new El Dorado opened in 2013. It is both a symbol of and a route to Colombia's growing competitiveness. Improved air connectedness through efficient and modern airport infrastructure is crucial for two booming sectors in the Colombian economy: cut flowers and tourism.

With an official slogan of 'Colombia, the only risk is wanting to stay!', not only does the new airport welcome international travellers to the country, it is evidence of the country's open-door commitment to newly established trade agreements with the US and Europe—not to mention evidence of a range of new foreign investors. The upgraded infrastructure around the airport services a competitive cargo trade, at the centre of which is floriculture, an industry that is deeply reliant on high standards of efficient transport to and from global markets.

Wider implications

Five key pointers for other nations fighting insurgencies on their own territory emerge from the study of Colombia's transformation over the past decade and a half.

First, the politics have to be right. What is striking about the Colombian experience is just how long it took to get into the position of electing a president with a firm democratic mandate to be tough on security. Since 1964, the natural inclination had been to attempt to contain the security problem while trying to achieve a negotiated peace. Yet over this long period the situation got slowly worse. In Colombia until the late 1990s the insurgency was seen as a 'war', in the words of a senior Air Force commander, 'between the guerrillas and the army'.[7] Only when the peace negotiations attempted by President Andres Pastrana failed and the country was on the verge of collapse was the democratic case firmly made for a radical change of course. In simple terms, as the guerrillas started to affect the lives not only of those in the countryside but also in the cities, including the capital, the war went from being someone else's problem to being everyone's business. This put improving security at the heart of the government's agenda and ensured both strong leadership and a cross-governmental approach; such an approach is vital for success for there is no peace without security. Political leadership has to remain flexible and agile in response to changing circumstances. President Uribe's election in 2002 and 2006 and President Santos's election in 2010 were votes for security. Today, Colombia's political debate has shifted to peace talks in Havana, and President Juan Manuel Santos's re-election for another four years in June 2014 was interpreted as a vote for peace.

Second, to run a successful COIN campaign you need, in the words of Field Marshal Montgomery regarding Malaya, both a 'man with a mandate' and a 'plan'.

While his attempts at peace negotiations ultimately failed, what President Andres Pastrana's administration started with Plan Colombia laid the foundations for the transformation of the military and the conduct of a successful COIN campaign. Successive administrations have built on and adjusted the plan, but they have all stuck with its overall approach. When trends began to reverse in 2010, Defence Minister Pinzón directed an in-depth review of the campaign, which resulted in the Sword of Honour and Green Heart plans that together put things back on track. These new plans, which show a clear lineage from Plan Colombia, have to date been comprehensively reviewed three times. It has been a process of continuous improvement, not radical shifts in direction. This ownership of the plan and continuity of approach has been a key element in the improving security picture. The contrast with the campaign in Afghanistan—where President Hamid Karzai had a bewildering array of plans and approaches to choose from, none of which were owned and driven by his government—is stark.

Indeed, as we have argued, this is one of the most significant differences between 'third-country counterinsurgency campaigns' such as in Iraq, Afghanistan or Vietnam—where an external intervener attempts to defeat an insurgency by deploying large expeditionary forces that lack local knowledge and tend to side-line and disrupt the local government—and situations like Indonesia, Colombia, Sri Lanka or El Salvador, where local government retains control and direction and carries the main burden of combat operations, drawing on outsiders only for technical advice, funding and limited support. Taking a broad view, it's clear that this difference—which we might colloquially shorthand as the 'home team advantage'—is a key predictor of success in COIN campaigns.[8]

In the case of Colombia, even with such a home-team advantage, security efforts have required more personnel, equipment, and money for the armed forces, which now number (with the police) nearly half a million personnel. The armed forces have made pragmatic new equipment acquisitions, befitting Colombians' highly practical, no-nonsense outlook. The air force, for example (as explained in Chapter 2) has focused on turboprop attack aircraft, and attack and transport helicopters, rather than flashy fighter-jets. But better security demands more than money or equipment. It is about better intelligence, improved inter-service cooperation, clever tactics, good training, high morale, and a sense of common mission and purpose. All of these attributes are now clearly visible.

The cost of the conflict has been heavy in treasure, blood and personal commitment, with more than 15,000 government casualties since 2000. Some troops whom the authors observed in the field during operations around La Macarena—the historic epicentre of the COIN campaign to the south-east of the capital—had been on station for eleven years, visiting their family one month in every six.

Fundamentally, the best tool for security is recruiting good people and using them well. The Colombians have made a real investment in the long-term development of their security force personnel and it shows.

Third, although—as noted—outsiders cannot run the show, they can help if they provide carefully targeted assistance under local government direction. International cooperation, such as with the US government under Plan Colombia from 2000, has been vital to Colombia's success. It has worked because the US and other nations have provided what the Colombians have wanted, not just what they had immediately available. Assistance has been carefully focused to fill gaps in knowledge, skills and equipment and—critically—these gaps were those identified and prioritised by Colombian planners and strategic leaders themselves. Throughout, there has been a ruthless focus by these Colombian leaders on acquiring what is needed for the fight, on getting bang for buck, and on retaining sovereignty. In the end, international will cannot substitute for domestic will. Now, the Colombians are cooperating extensively across South America and the Caribbean, themselves offering training in the fight against drugs.

Fourth, offering an alternative to the political economy of war is vital to undermine those conflict entrepreneurs who seek to perpetuate wars for their own narrow financial interests. Yet re-establishing a 'normal', legal economy in the rural areas is, as the Colombians have found, very difficult—not least since the guerrillas have worked hard to keep these populations dependent on their preferred alternative and since now, fifty years on, many people know little different or better than to live off a cocktail of crime and coca. To achieve such an effect the government needs a tangible, effective, presence in the remote rural areas; this task cannot be left to the military.

Alternatives have to be provided and nurtured. Geographically there is a stark correlation between levels of poverty and the presence of the insurgency. By 2014, Colombia had reduced its area under drug cultivation to just a third of the level of 2000 and made some progress in providing alternative livelihoods, but there remains much more to be done. The 2015 coca cultivation figures indicate a disappointing increase in areas where the government had suspended spraying, illustrating how fragile the gains are, and how easily they can be reversed. This is an issue that has considerable resonance in Africa, particularly with its existing economic starting point, predicted rapid population growth and urbanisation to match, but lack of commensurate job creation. As noted in the Introduction, Congolese and Central African Republic warlords, for example, have undergone a commercial metamorphosis similar to that of FARC, from guerrilla to conflict entrepreneur, greatly lengthening the instability suffered by their countries.

All this relates to the fifth and final pointer from Colombia: change takes time and demands continuity of approach. For example, with two-thirds of Buenaventura's population unemployed, high levels of illiteracy, a virtually non-existent culture of education, and jobs unsurprisingly few and far between, systemic change will take perhaps a generation. A sense of alienation, hopelessness and a lack of personal responsibility, especially of young males, is deeply ingrained within poorer elements of Colombian society.

Exiting a cycle of conflict and violence demands the provision of a peaceful, stable environment for young people, so that future generations can build on the achievements of the present. It is about delivering a whole-of-country approach that brings government to once ungoverned spaces, removing the spaces where violence and lawlessness can flourish.

Prospects for peace

So with all this progress, why do tens of thousands of people still support FARC?[9] Are these just people who are stuck in the past, whose families have taken sides and suffered casualties, and as a result are unable to move forward? Or do the deep divisions in society that gave birth to FARC and the other guerrilla movements still remain, set to re-emerge after the security crisis is over?

The truth is that both of these explanations are true. Dealing with the physical and emotional consequences of the war will be take generations. Even a quick study of the Balkan wars, the conflict in Northern Ireland, or indeed the American Civil War, for example, reveals this to be the case. The reduction of inequalities through the rebalancing of political power and wealth will take just as long. Sustainable economic development and job creation does not happen overnight—and if the political and governance focus was to quickly shift from the rural areas back to the cities where the vast majority of Colombians reside it could easily fan the embers of conflict. Success will be a function of the need to deliver increasing wealth through economic growth alongside a more equitable distribution of that wealth. This involves a process of nationbuilding as much as it does the securing of peace.

Success will also reflect the ability of both the government and FARC (along with its ELN and EPL guerrilla allies) to reform and transform. In this regard, the recent history of El Salvador is thought-provoking. A UN-brokered peace agreement in 1992 ended the twelve-year-long civil war; the Chapultepec Peace Accords mandated reform in the armed forces and police, and the government agreed to submit to the recommendations of a Commission on the Truth for El Salvador. In 1993, the commission delivered its findings and the legislature quickly passed an amnesty law covering acts of violence committed in the war. Economic reforms since the war have improved living conditions to an extent, but high crime rates and gang culture remain significant obstacles to external investment and land reform remains a contentious issue. For seventeen years following the conflict, Salvadorans continued to vote for the right-wing National Republican Alliance party (ARENA)—the party that had delivered peace. In 2001, as a result of frustrations at the slow pace of economic progress and internal divisions within ARENA, the Farabundo

Martí National Liberation Front (FMLN) party began to make electoral gains. Then, in 2009, in another break with the past, the FMLN elected Mauricio Funes—a journalist rather than a former guerrilla—as its leader. In the following presidential election, and on the back of widespread public dissatisfaction with the corruption of the incumbent administration, Funes was elected as El Salvador's first FMLN president of the country, thus completing the party's integration into democratic politics.

In Colombia, with the election of Gustavo Petro—a former member of the M-19 guerrilla group—as first a senator, and then as mayor of Bogotá in 2012, Colombians have already experimented with a move in this direction. There is widespread debate about the outcome, fuelled in part by his temporary suspension as mayor in 2014. The issue for Colombia is whether FARC, having rejected peaceful politics, can reinvent itself as a political party following a peace process and play a meaningful role in the democratic political life of the country—or indeed, whether FARC leaders have any intention of following this path.

As we argued in Chapter 5, there is in fact strong evidence that FARC does not intend to transform itself into a purely legal political party. FARC—drawing ideas from diverse sources like Gene Sharp's work on democratic revolution and the experience of the Movimiento al Socialismo (MAS) and the Cocaleros in Bolivia—may have something different in mind. It may be planning to apply its proven strategy—the combination of all forms of struggle—to continue its effort to seize power even after a peace deal. This would be by means of carefully calibrated violence, illicit funding (through continued drug trafficking and extortion), manipulation of social protest, formation of licit political parties to cement gains achieved through illegal means, international propaganda, and efforts to neutralise and destroy the military. There is a risk of a Bolivia-style outcome, and given the lopsided commitments given by each side at the peace table in Havana to date, there is a real imperative for Colombian leaders—and for the Colombian people—to remain extraordinarily vigilant, lest what has been won on the ground since 1999 be traded away at the negotiating table. In this respect a mooted ceasefire, as a prelude to a wider peace deal, contains some of the risks that undermined Pastrana's peace process in 1998, where FARC took advantage of government concessions to strengthen regional control and to rearm and reequip.

For its part, the government, too, has to remain true to its rhetorical and policy commitments on social justice and inclusive growth, without which historical inequalities will persist, within the cities, between the cities and the rural areas, and between the *campesinos* (peasants) and the landed elites.

The path of progress

Assuming that the peace process can be managed in a way that avoids this negative outcome and results in real and sustainable peace—admittedly, a huge assumption—what does the evidence suggest will be needed in order to achieve continued progress?

First, an improved security situation will need not only to be maintained, but further enhanced. The peace process is dependent on improved security and confidence. Sustaining this momentum will not be easy; for many, drugs and crime have become a way of life. For as long as FARC refuses to surrender its weapons, and BACRIM and the ELN remain in the field, the risk of renewed violence will be ever-present. In this respect, early disbandment of Joint Task Forces (being considered by the government as of mid-2015) may paradoxically undermine the very peace process it is designed to deliver. The same applies to calls for the defence budget to be slashed in the face of a cease-fire and wider economic pressures. Managing the tension between the need to insure against a return of insecurity and the resource cost to the nation of doing so will require adroit political juggling and exceptionally fine judgment calls.

Second, whatever is finally agreed at the peace table needs to be delivered in a manner that is transparent and verifiable, and that holds both the government and former insurgents to account on real commitments. The task of rebuilding trust and confidence, after decades of fighting, will not be easy.

Third, and much more challenging, the real drivers of the conflict—economic and political exclusion, the mismatch between rural and urban areas, and the two-speed economy—must continue to be visibly tackled. Not only are the capabilities and finances of the military itself subject to the same overall national financial constraints, but the wider goal of a more inclusive growth agenda is likely to be complicated by any economic downturn.

Ultimately, if the people of Colombia buy into their political and economic systems with just a small fraction of the enthusiasm with which they support their national football team, then the conflict will recede. This will require the delivery of a demonstrable rate and direction of change to the Colombian people. This will be difficult to achieve, however, if global commodity prices, to which the Colombian economy and its currency is vulnerable, remain downwardly volatile, complicating, too, any achievement of a financial peace dividend.

The challenges of the peace process for the military and the guerrillas are equally huge; for the consequence for both is disarmament and outright or partial demobilisation. Ultimately, it will entail—in different ways for each—a loss of influence, income and, in the longer-term, prestige. As a result, some of the less ideologically motivated guerrillas will undoubtedly migrate into the criminal sphere, whereas some dissident elements—conflict entrepreneurs—will probably continue to engage in criminal activity in order to support the continuing combination of all forms of struggle. For the military, the challenge will be to continue to support the police in the delivery of security; to transform into a force focused on the next generation of threats (whether internal or external); and to all the while retain operationally experienced, talented people. Furthermore, the large numbers of demobilised armed forces personnel will need to be accommodated in the civilian economy, lest they, too, become part of the security problem. This task may be eased by the huge amount Colombia's armed forces have to offer in the international arena—as part of peacekeeping missions, or in training and advising others.

For both the government and FARC there is a perfectly logical—but highly immoral—argument for keeping a low-intensity fight going. For adjusting the balance of power as part of a negotiated settlement will be painful for all. There can, for example, be no equivalence between the armed forces of a democratically elected government lawfully conducting their duties, and narco-guerrillas operating outside the law. Furthermore, if a peace deal is reached and the guerrillas lay down their arms, the financial costs of delivering a verifiable peace and its associated reforms are potentially very high. Such an argument is, however, unlikely to gain traction.

In the background, too, there is the added tension of time. For President Santos, in particular, elected on a mandate for peace and unable to stand for a third term, the electoral clock runs out in mid-2018. For FARC, on the other hand, fighting since 1964, the only motivation to move quickly to a deal is likely to be continued military pressure—and this has been reduced through negotiation, as has drug eradication—as well as any variation in external sanctuary and support. To the average Colombian, a lengthy negotiation process, with no end in sight, might demonstrate a lack of serious intent by FARC. Time will tell.

Even after fifteen years of tough fighting and government success, sustainable peace remains some way off. Wrong policy and resource choices could easily unravel the hard-won security and governance gains.

The overall lesson to take from Colombia's progress towards stability is clear: get the security piece right with a firm cross-government approach, enabling the consolidation of institutions, and much else follows. As we explain in our analysis of the contrast between counter-guerrilla successes and broader COIN challenges in Colombia, we consider that security is too fundamental an issue to be left simply to the military and police. In any case, the security forces lack the means to effect—and, in a democracy, are rightly forbidden from meddling in—the political solutions needed to deal with the underlying causes of the conflict. It is a lesson that is highly relevant to those currently battling, or attempting to pre-empt, insurgencies in Africa.

Leadership is the key to achieving such solutions. A good leader's attention to detail in executing a vision and a plan enables such a cross-government approach, gets the best out of talented people, and allows difficult policy choices to be made. Sustained commitment in the face of adversity is equally vital, as experience shows that bringing insurgencies to an end takes much more time, blood, and treasure than many leaders will readily acknowledge.

NOTES

PREFACE

1. Fidel Castro, *La Paz en Colombia*, Editora Politica, 2009.

INTRODUCTION: COLOMBIA'S TRANSITION

1. See Steven Dudley, *Walking Ghosts: Murder and Guerrilla Politics in Colombia*, New York: Routledge, 2006, p. 2.

2. These 'independent republics' included Agriari, Viota, Tequendama, Sumapaz, El Pato, Guayabero, Suroeste del Tolima, Rio Chiquito, 26 de Septiembre and Marquetalia.

3. See Dennis M. Rempe, 'Guerrillas, Bandits, and Independent Republics: US Counter-insurgency Efforts in Colombia 1959–1965' in *Small Wars and Insurgencies*, Vol. 6, No. 3 (Winter 1995), pp. 304–327.

4. Ibid.

5. Ibid.

6. Quoted in James Monahan and Kenneth O. Gilmore, *The Great Deception*, New York: 1963, p. 153.

7. For a description of Cuban and Soviet support for regional guerrilla movements (and its limitations), see Jay Mallin, 'Phases of Subversion: The Castro Drive on Latin America', *Air University Review*, Vol. 25, No. 1 (November–December 1973), pp. 54–62.

8. Police Intelligence Division, 10 December 2010.

9. In conversation with a delegation of visiting African leaders, 19 June 2014.

10. Espen Barth Eide, '"Conflict Entrepreneurship": On the "Art" of Waging Civil War' in Anthony McDermott (ed.), *Humanitarian Force*, Oslo: International Peace Research Institute, 1997, pp. 41–69.

11. United States Institute of Peace, *Glossary of Terms for Conflict Management and Peacebuilding*, online at http://glossary.usip.org/resource/conflict-entrepreneur, last accessed 17 August 2015.

12. The term 'conflict entrepreneur' was coined by Espen Barth Eide in 1997 to describe a class of actors in civil wars and intra-state conflicts who deliberately initiate conflicts to further their personal interests. We use the term slightly differently here, to describe those who perpetuate or prolong a conflict for the same reasons, or for motivations more closely related to plunder and profit than to political goals per se. See Espen Barth Eide, 'Conflict Entrepreneurship: On the Art of Waging Civil War' in McDermott (ed.), *Humanitarian Force*, pp. 41–70.

13. At Puerres in the department of Nariño on 5 April 1996, resulting in thirty-one government troops killed; Las Delicias in Putumayo on 30 August 1996, with thirty-one killed and sixty wounded; La Carpa in Guaviare on 6 September 1996, with twenty-three dead; Patascoy in Guaviare on 21 December 1997, with twenty-two killed and eighteen wounded; El Billar in Caquetá on 5 May 1998, with sixty-three dead and forty-three wounded; Miroflores in Guaviare on 3 August 1998, with nine dead and twenty-two wounded; Mitú in Vaupes on 1 November 1998, with thirty-seven dead and sixty-one wounded; and Jurado in the department of Chocó on 12 December 1999, with twenty-four government troops killed.

14. Mills, discussion with Colombian respondents, Bogotá, 9 December 2010.

15. Plan Colombia was a comprehensive whole-of-government plan to stabilise the country developed by the Colombian administration with significant advice from the US government under President Bill Clinton. In support of Plan Colombia, the US Congress approved an initial $1.1 billion aid package for counter-narcotics activities. This aid package is often erroneously thought of as Plan Colombia.

16. Oscar Naranjo, 'Colombia Shows the Value of Cooperation', *New York Times*, 17 April 2013.

17. In the 1960s FARC termed their strategy: *La combinación de todas la formas de lucha* (the combination of all forms of the struggle). While they had left-wing politicians, unionists, students and others representing their interests in formal chambers, the guerrillas took the fight to the government in the mountains and hills. Dudley, *Walking Ghosts*, p. 8.

18. Mills, discussion with representatives from Ministerio de Defensa Nacional Colombia, 2014.

19. Operational Design, a technique pioneered by planners in the United States in the early twenty-first century, uses a combination of systems analysis and design thinking to model both friendly and adversary structures and systems, deriving operational effects from an analysis of vulnerabilities and critical requirements of all parties to a conflict. Colombian officers trained in these methods led critical elements of the Sword of Honour planning program. Kilcullen: author's participant observation during the Sword of Honour planning process, 2011–12.

20. Mills: discussion with General Rodriguez, Bogotá, 2013.

21. Brigadier General Salgado, *National Army Transformation Brief*, Ministry of Defence of the Republic of Colombia, February 2015.

22. See Milburn Line, 'Trying to End Colombia's Battle With FARC', *Foreign Affairs*, 27 March 2012.
23. See Meredith Reid Sarkees and Frank Wayman, *Resort to War: 1816—2007*, Washington, DC: CQ Press, 2010.
24. For a detailed intellectual history of counterinsurgency, including an analysis of the classical 'canon' of cases and its limitations, see David J. Kilcullen, 'Counterinsurgency: The State of a Controversial Art' in Paul B. Rich and I. Duyvesteyn, *The Routledge Handbook of Insurgency and Counterinsurgency*, London: Routledge, 2012.

1. A LONG WAR

1. As explained in David Spencer et al, *Colombia's Road to Recovery: Security and Governance 1982–2010*, Washington, DC: National Defense University, 2012.
2. Centro Nacional de Memoria Historica, *Guerrilla y población civil: Trayectoria de las FARC 1949–2013*, Centro Nacional de Memoria Historica, 2014, pp. 16–20. According to this study phase 1 runs from 1949–78, phase 2 from 1979–1991, phase 3 from 1991–2008 and phase 4 from 2008–2013.
3. Dennis M. Rempe, 'Guerrillas, Bandits, and Independent Republics: US Counterinsurgency Efforts in Colombia 1959–1965', *Small Wars and Insurgencies*, Vol. 6, No. 3 (Winter 1995), pp. 304–327.
4. FARC, *Segunda Conferencia de las FARC*, 1966.
5. Daniel Pecaut, *Las FARC: Una Guerrilla Sin Fin o Sin Fines*, Bogotá: Editorial Norma, 2008, p. 90.
6. Ejército Nacional de Colombia, *Evolución de las FARC 1964–2005*, 2005.
7. FARC-EP, *Conclusiones de la Septima Conferencia*, May 1982.
8. UP won 350,000 or 4.5 per cent of the popular vote for their presidential candidate; they also won two senate seats and ten representative seats, and gained twenty-three mayors and 351 councilmen. Probably one of the best English language sources on the UP is Steven Dudley, *Walking Ghosts*, New York: Routledge, 2004.
9. Ministro de Gobierno Fernando Cepeda Ulloa, *Memoria al Congreso*, 7 August 1986–17 August 1987, p. 157.
10. FARC-EP, *Conclusiones de la Septima Conferencia*, May 1982.
11. Ibid.
12. FARC-EP, *Conclusiones Generales del Pleno del Estado Mayor de las FARC*, May 1989.
13. Ibid.
14. A good example is found in Russell Stendall, *Rescue the Captors*, Ransom Press International, 1986.
15. For example see 'Genocidio partido político Unión Patriótica', memoriaydignidad.org, http://memoriaydignidad.org/memoriaydignidad/index.php/casos-

emblematicos/141-masacres-1980-a-2010/640-genocidio-de-la-up, last accessed 24 August 2015.

16. Ministro de Gobierno Fernando Cepeda Ulloa, *Memoria al Congreso Nacional*, 7 August 1986–17 August 1987, p. 17.

17. For example, see 'Las FARC en las Goteras de Bogotá', *El Tiempo*, 6 February 1991.

18. 'Ultimo Parte de Guerra de 1991', *El Tiempo*, 9 February 1992.

19. 'Regresa a Comision Negociadora', *El Tiempo*, 22 March 1992.

20. FARC-EP, *Informe de la Octava Conferencia de las FARC-EP Comandante Jacobo Arenas Estamos Cumpliendo*, April 1993.

21. Ibid.

22. Ibid.

23. Ibid.

24. Ibid.

25. Ibid.

26. Ibid.

27. Ibid.

28. Ibid.

29. FARC-EP, *Pleno del Estado Mayor Central*, November 1997.

30. Ibid.

31. Glenn E. Curtis and Tara Karacan, The *Nexus among Terrorists, Narcotics Traffickers, Weapons Proliferators, and Organized Crime Networks in Western Europe*, Washington, DC: Library of Congress Federal Research Division, 2002, p. 6.

32. Edgard Tellez, Oscar Montes and Jorge Lesmes, *Diario Intimo de un Fracaso: Historia no Contada del Proceso de Paz con las FARC*, Editorial Planeta, 2002, p. 51.

33. 'Asi Registro El Tiempo la Toma de Mitu, Vaupes (3 de Noviembre de 1998)', *El Tiempo*, 16 March 2010.

34. FARC-EP, *Pleno del Estado Mayor Central: 'Con Bolívar, por la paz y la soberanía nacional'*, March 2000.

35. Ibid.

36. Ibid.

37. Ibid.

38. National Police of Colombia, *Cultivos de Coca en Colombia 1995–2005*, March 2006.

39. United Nations Office of Drugs and Crime, 'Sistema Integrado de Monitoreo de Cultivos Ilícitos', http://www.unodc.org/colombia/es/simci2013/simci.html, last accessed 6 August 2015.

40. FARC leader Mono Jojoy said: 'La Guerrilla va para las ciudades. Alla nos pillamos.' (The Guerrillas are going to the cities. We'll run into you there.) See 'FARC anuncia Guerra Urbana', *El Tiempo*, 25 June 2001.

41. '30 Militares Muertos en Ataque de las FARC, *El Tiempo*, 23 June 2001.

42. See Claudia Rocio Vasquez, 'Retirada Guerrillera Hacia el Caguan', *El Tiempo*,

22 August 2001; see also 'Accoraladas las FARC en Mapiripan', *El Tiempo*, 24 August 2001.

43. Secretariado del Estado Mayor Central FARC-EP, Pleno Extraordinario, January 2002.

44. Ibid.

45. Juan Forero, 'Colombian Rebels Hijack a Plane and Kidnap a Senator', *New York Times*, 21 February 2002.

46. Manuel Marulanda, *Ponencia para el Desarrollo de la Novena Conferencia Nacional Guerrillera*, 2007.

47. Directive from Jorge to Aurelio dated 4 February 2002. (Document captured by the Colombian Army.)

48. Manuel Marulanda, *Circular del Comandante Manuel*, 2002.

49. Directive from Jorge Suarez Briceño to Marco Aurelio Buendia, Hugo, Geovanny, Pablo and Javier, 17 July 2002. (Document captured by the Colombian Army.)

50. Ibid.

51. FARC-EP, *Informe para Balance de los Frentes Policarpa Salvatierra y 42 del Bloque Oriental de las FARC-EP*, 15 March 2004.

52. An area comprising a rough square between the Sierra de la Macarena in the north-west, San Jose del Guaviare in the north-east, the Serrania de Chiribiquete in the south-east, and Peñas Coloradas in the south-west.

53. Spencer: interview with anonymous deserter, 14ᵗʰ Front, deserted in January 2006.

54. Manuel Marulanda, *Ponencia para el Desarrollo de la Novena Conferencia Nacional Guerrillera*, 2007.

55. Ibid.

56. The most thorough account of this operation is found in Juan Carlos Torres, *Operacion Jaque: La Verdadera Historia*, Bogotá: Planeta, 2008.

57. 'Captured Raul Reyes documents', released by the Colombian government in April 2008. For a more in-depth analysis see Nigel Inkster and James Lockhart Smith, *The FARC Files: Venezuela, Ecuador and the Secret Archive of 'Raúl Reyes'*, London: International Institute of Strategic Studies, 2010.

58. Jose Gregorio Perez, Raul Reyes, *El Canciller de la Montaña*, Bogotá: Grupo Editorial Norma, 2008, pp. 185–250.

59. Alfonso Cano, 'Plan Renacer Revolucionario de las Masas', 16 August 2008.

60. Ibid.

61. Ibid.

62. Ibid.

63. National Police of Colombia Crime Statistics 2002–2010.

64. Ibid.

2. BUILDING THE TOOLS FOR MILITARY SUCCESS

1. Anthony Arnott is a former officer and pilot in the British Army Air Corps, and a Master's graduate of Singapore's Rajaratnam School of International Studies. He divides his time between the Brenthurst Foundation and E Oppenheimer & Son and visited Colombia with the Brenthurst group in June 2015.

2. Álvaro Uribe Vélez, *No Lost Causes*, New York: Penguin, 2012, p. 166.

3. Calculated as gross assets minus debt. The calculation excluded the first $117,000 of a main residence.

4. For more information see Gustavo A. Flores-Macías, 'Financing Security Through Elite Taxation: The Case of Colombia's "Democratic Security Taxes"', International Centre for Tax and Development, Working Paper 3, July 2012, www.ictd.ac/sites/default/files/ICTD_WP3.pdf, last accessed August 2015.

5. Vélez, *No Lost Causes*, p. 167.

6. Arnott and Davis: interview with General Jorge Hernando Nieto Rojas, Bogota, March 2015.

7. Jim Rochlin, 'Plan Colombia and the revolution in military affairs: the demise of the FARC', *Review of International Studies*, 37(02), 2011, pp. 715–740.

8. Adriaan Alsema, 'DAS should not be disbanded: Presidency', *Colombia Reports*, 25 February 2009.

9. All interviews in support of this section on intelligence were conducted in March 2015.

10. For example: 'Covert action in Colombia', *Washington Post*, 21 December 2013, http://www.washingtonpost.com/sf/investigative/2013/12/21/covert-action-in-colombia/, last accessed 10 August 2015.

11. Álvaro Uribe Vélez, *No Lost Causes*, pp. 172–3.

12. The Colombian relationship with Israel is a long one with diplomatic relations reaching back more than fifty years: a battalion of Colombian soldiers has been committed to the Multinational Force and Observers in the Sinai since 1980 and there have been considerable equipment sales.

13. Arnott: interviews with Colombian Air Force personnel, March 2015.

14. 'Covert action in Colombia', *Washington Post*, 21 December 2013, http://www.washingtonpost.com/sf/investigative/2013/12/21/covert-action-in-colombia/, last accessed 10 August 2015.

15. *Ibid.*

16. Davis and Kilcullen: interviews with Special Forces personnel, Bogota, March 2015.

17. According to UK doctrine the eight defence lines of development are: Training, Equipment, People, Information, Doctrine, Organisation, Infrastructure and Logistics. See: www.gov.uk/government/uploads/system/uploads/attachment_data/file/36720/20090210_MODAFDLODAnalysis_V1_0_U.pdf, last accessed August 2015.

18. Davis: interview with Colombian defence official, Bogota, March 2015.

3. GUERRILLA AND COUNTER-GUERRILLA WARFARE IN COLOMBIA

1. Unless otherwise referenced, insights in this chapter are drawn from interviews and participant observation field-notes gathered during Kilcullen's fieldwork in Colombia during 2009, 2011, 2013, 2014 and 2015, and his participation in the external review team for the revised Sword of Honour II plan in 2012, as well as interviews with demobilised FARC guerrillas in June and September 2014.

2. United States Department of Defense, Joint Publication 3–24, *Counterinsurgency*, Washington DC: November 2013, pp. I-2.

3. United States Army, Field Manual 90–8, *Counterguerrilla Operations*, Washington, D.C.: May 1986, pp. 1–5.

4. The term 'counter-guerrilla' also sometimes denotes irregular anti-guerrilla (e.g. paramilitary or militia) forces that emerge spontaneously to fight insurgents, and exist outside the framework of (and sometimes in competition with) legal state institutions—for example, Loyalist paramilitaries in Northern Ireland, Montagnard guerrillas in Indochina, Colombia's *Autodefensas* or the Iraqi popular mobilisation forces. This is not how the term is used here: rather it simply denotes the subset of an overall COIN campaign that is primarily the responsibility of the security forces.

5. For a discussion of FARC's origins in the independent republics of the post-Violencia period, as well as the role of United States counterinsurgency assistance in this process see Rempe, 'Guerrillas, Bandits, and Independent Republics: US Counterinsurgency Efforts in Colombia 1959–1965', *Small Wars and Insurgencies*, Vol. 6, No. 3 (Winter 1995), pp. 304–327.

6. David Galula, *Counterinsurgency Warfare: Theory and Practice*, London: Pall Mall, 1964 pp. 35–8

7. The following countries directly border Colombia: Brazil (with a border length of 1,644 km), Ecuador (590 km), Panama (225 km), Peru (1,800 km) and Venezuela (2,050 km). These countries—with the exception of Venezuela—have not historically offered direct support to FARC, but the ease of border crossing, the availability of jungle and mountain sanctuaries, and the inability of several of Colombia's neighbours to control their frontier regions have offered significant advantages to FARC.

8. In order, the four largest countries in South America by territorial extent are Brazil (with an area of 8,514,877 square kilometres), Argentina (2,766,890), Peru (1,285,220) and Colombia (1,138,910), while the largest by population are Brazil (with 191.2 million people as of July 2009), Colombia (45.9 million), Argentina (40.5) and Venezuela (31.6).

9. Philip Kelley, *Checkerboards and Shatterbelts: The Geopolitics of South America*, Austin, TX: University of Texas Press, 2010, p. 78.

10. James D. Henderson, *Colombia's Narcotics Nightmare: How the drug trade destroyed peace*, Jefferson, NC: McFarland, 2015, p. 16.

11. Readers unfamiliar with guerrilla warfare may think of jungles, mountain ranges

and rivers as terrain compartments that constrain movement. This is true for conventional, mechanised forces like those of the United States or many European powers. For guerrillas—light infantry forces, moving primarily on foot and dispersed in small groups, with a relatively small logistics footprint and the ability (for the most part) to set the tempo of their own operations—such terrain features act as mobility corridors allowing the guerrilla to transit large distances without detection.

12. Henderson, *Colombia's Narcotics Nightmare*, p. 122.
13. According to the Oxford English Dictionary, the first use of the term 'narco-terrorism' was by the then President of Peru, Fernando Belaunde Terry, in 1982. See *Oxford English Dictionary*, entry for 'narco-, comb. form', at http://www.oed.com/view/Entry/125094?rskey=PPblkP&result=2#eid, last accessed 11 August 2015.
14. Smedley D. Butler, *War is a Racket*, New York: Feral books, 2003 (first published 1935).
15. Henderson, *Colombia's Narcotics Nightmare*, pp. 134–5.
16. See LaVerle Berry, Glenn E. Curtis, Rex A. Hudson and Nina A. Kollars, *A Global Overview Of Narcotics-Funded Terrorist And Other Extremist Groups* (2002) Washington, D.C.: Library of Congress, May 2002, pp. 50 ff.
17. Thomas R. Cook, 'The Financial Arm of the FARC: A Threat Finance Perspective' in *Journal of Strategic Security*, Volume IV, Issue 1, 2011, p. 21.
18. Henderson, *Colombia's Narcotics Nightmare*, p. 157.
19. Elyssa Pachico, '70% of FARC Assets Held Outside Colombia', Insight Crime, 18 September 2012, at http://www.insightcrime.org/news-analysis/70-of-farc-assets-held-outside-colombia, last accessed 11 August 2015.
20. Ibid.
21. See Jeremy McDermott, 'Is Colombia Again the World's Top Cocaine Producer?', Insight Crime, 6 May 2015, online at http://www.insightcrime.org/news-analysis/colombia-again-world-top-cocaine-producer, last accessed 11 August 2015.
22. Fidel Castro Ruz, *La Paz en Colombia*, Havana: Editora Política, 2008.
23. Berry et al., *A Global Overview*, p. 51.
24. See 'Exodus of FARC leaders to Cuba drains rebels' military command', *Colombia Reports*, 27 October 2014, online at http://colombiareports.com/farc-placing-high-level-commanders-cuba-peace-talks-shows-commitment-peace-talks/, last accessed 12 August 2015.
25. See International Institute for Strategic Studies, *The FARC Files: Venezuela, Ecuador, and the Secret Archive of Raúl Reyes*, London: IISS, 2011.
26. Henderson, *Colombia's Narcotics Nightmare*, p. 104.
27. See for example Peter J. Moons, *The Colombian Miracle: How the Government of Colombia Beat Back an Insurgency and Saved the Nation*, 2013.
28. Insurgencies can last between fifteen and thirty years, while reconstruction often takes far longer. For example, the Malayan Emergency began in June 1948 and the

insurgents were effectively neutralised by 1959. However, it took another thirty years—until 1989—for Communist Party leaders to surrender. During that time the Royal Malaysian Armed Forces (with assistance from Australia, New Zealand and the United Kingdom) maintained an internal security role, while the government undertook a national development plan to address the grievances that had driven the insurgency. Likewise, a US Defense Science Board summer study in 2004 found that stabilisation and reconstruction operations undertaken by the United States since the end of the Cold War lasted, on average, ten years. See United States Department of Defense, *Defense Science Board Summer Study 2004: Transition to and From Hostilities*, Washington D.C.: Office of the Undersecretary for Acquisition, Technology and Logistics, 2005, pp. 6–10, online at http://www. acq.osd.mil/dsb/reports/ADA430116.pdf, accessed 17 August 2015.

29. Kilcullen: discussion with a Colombian strategic analyst, Tolemaida, August 2014.

30. Kilcullen: interview at forward operating base with troops of JTF Nudo de Paramillo, 12th June 2014.

4. THE DOOR THROUGH WHICH MUCH FOLLOWS? SECURITY AND COLOMBIA'S ECONOMIC TRANSFORMATION

1. Grateful thanks are expressed to Anthony Arnott and Dr Lyal White for their help in the preparation of this chapter, which is based on research undertaken in Cartagena and Medellín in December 2006, and again in Colombia in September 2008, November–December 2010, November 2013, June–July 2014 and March 2015.

2. This data is drawn from Jorge Giraldo-Ramírez and Andrés Preciado-Restrepo, 'Medellín, from Theater of War to Security Laboratory', *Stability: International Journal of Security and Development*, June 2015 at http://www.stabilityjournal. org/articles/10.5334/sta.fy/, last accessed 25 August 2015.

3. Francis Fukuyama and Seth Colby, 'Half a Miracle', *Foreign Policy*, 25 April 2011, at http://foreignpolicy.com/2011/04/25/half-a-miracle/, last accessed 25 August 2015.

4. Ibid.

5. Data supplied by Fedesarollo, March 2015.

6. Mills: discussion, Police Intelligence, Bogotá, March 2015.

7. Small criminal structures, from gangs to common criminals, that are typically involved in extortion and micro-trafficking business, often servicing other larger organised crime structures.

8. Mills: discussion, Medellin, March 2015.

9. The population of Antioquia totals 6 million people. Of this number, approximately 2,500,000 inhabitants are from Medellín. The inhabitants in Medellín and the wider metropolitan area together add up to 3,500,000.

10. By comparison, in 2000 there were 5,863 police officers across in the metropoli-

tan area. Information supplied in correspondence by Anamaría Botero Mora, Agencia de Cooperación e Inversión de Medellín y el Área metropolitan, 17 August 2015.

11. Giraldo-Ramírez and Preciado-Restrepo, 'Medellín, from Theater of War to Security Laboratory'.

12. The number of graduate officers has increased from 'around 10 per cent' in 1994 to '70 per cent of officers' in 2015, a result, primarily, of the 2004 human resource strategy. Mills: discussion, Medellin police, March 2015.

13. Mills: discussion, Medellín, March 2015.

14. For example, in June 2015, a commander of the Medellín crime syndicate 'Oficina de Envigado' said that its 'biggest ally' was the city's own police department. The Oficina, founded by Pablo Escobar, is the city's main criminal organisation, has a claimed 2000 members, and controls drug trafficking, drug dealing and extortion rackets. According to the commander, 'Our biggest ally here in Medellin is the police. Members of the police are receiving approximately 200 thousand to 300 thousand pesos [between $80 and $120] per week from us', or one-third of the monthly income of low-ranked officers. See http://colombiareports.com/medellin-police-are-citys-crime-syndicates-biggest-ally-capo/, last accessed 25 August 2015.

15. Giraldo-Ramírez and Preciado-Restrepo, 'Medellín, from Theater of War to Security Laboratory'

16. Mills: interview, November 2006.

17. Presentation, Fedesarollo, 18 March 2015.

18. Mills: discussion, Bogotá, March 2015.

19. See Matt Rendell, *Kings of the Mountain*, London: Aurum, 2002.

20. Mills: interview, Bogota, December 2006. See also Cardenas's 'Economic Growth in Colombia: A Reversal of "Fortune"?', Centre for International Development Working Paper No. 83, December 2001, http://www.hks.harvard.edu/content/download/69055/1249030/version/1/file/083.pdf, last accessed 13 August 2015.

21. Rendell, *Kings of the Mountain*, especially pp. 161–6.

22. Mills: discussion, US Embassy, Bogotá, March 2015.

23. Mills: discussion, Bogotá, 26 March 2015.

24. At Presidencia de la Republica-Ministerio de Defensa Nacional, *Politica de Defensa y Seguridad Democratica*. Cited in David Spencer et al., 'Finding a Solution: The Government of Alvaro Uribe, 2002–2010', in *Colombia's Road to Recovery: Security and Governance 1982–2010*, Washington, DC: Center for Hemispheric Defense Studies, National Defense University, June 2011, p. 63.

25. Mills: interview, Bogotá, December 2006.

26. Mills: interview, Bogotá, March 2015.

27. Mills: discussion, Jeronimo Uribe, Bogotá, March 2015.

28. See 'Passing the baton', *The Economist*, 31 July 2014, http://www.economist.com/news/

finance-and-economics/21610305-colombia-overtakes-peru-become-regions-fastest-growing-big-economy-passing, last accessed 13 August 2015.

29. See Cardenas, 'Economic Growth in Colombia'.
30. Mills: interview, Bogota, March 2015.
31. See 'Policy Brief 2', Overseas Development Institute, February 2006, http://www.odi.org/sites/odi.org.uk/files/odi-assets/publications-opinion-files/1690.pdf, last accessed 13 August 2015.
32. As of 2014 there are two types of incentives. First, families with children under seven years old, independent of the number of children, are paid monthly an amount between (peso) $61,200 (US$25) and $71,400 (US$28). Second, for each child between five and eighteen years old who is attending school, families receive between $25,000 (US$10) and $56,000 (US$22) per child (up to a maximum of three children per family). The family must ensure the attendance at school of their children. These figures were supplied by Fedesarollo, drawn from information supplied by Familias en Acción at http:///www.dps.gov.co. For further background on the grants, see Orazio Attanasio, Erich Battistin, Emla Fitzsimons, Alice Mesnard and Marcos Vera-Hernández, 'How Effective are Conditional Cash Transfers? Evidence from Colombia', Institute for Fiscal Studies, Briefing Note No. 54, 2005, at http://www.ifs.org.uk/bns/bn54.pdf, last accessed 25 August 2015.
33. Mills: discussion, November 2006.
34. Mills: telephonic discussion, Ken Kluksdahl, 26 March 2015.
35. Ibid.
36. Ibid.
37. See 'Latin America: Tax revenues are rising, but still low and varied among countries', OECD, http://www.oecd.org/chile/latinamericataxrevenuesarerisingbutstilllowandvariedamongcountries.htm, last accessed 13 August 2015; and 'Tax revenue (% of GDP)' indicator, The World Bank, http://data.worldbank.org/indicator/GC.TAX.TOTL.GD.ZS, last accessed 13 August 2015.
38. James Robinson, 'Colombia: Another 100 Years of Solitude?', Current History, February 2013, p. 44. See also his 'Orangutan in a Tuxedo', Legatum Institute Prosperity in Depth Series, 5, 2012.
39. Robinson, 'Colombia: Another 100 Years of Solitude?', p. 44.
40. Ibid.
41. 'Peace, land and bread', The Economist, 22 November 2012, http://www.economist.com/news/americas/21567087-hard-bargaining-starts-peace-land-and-bread, last accessed 13 August 2015.
42. GINI Index, (World Bank estimate), The World Bank, http://data.worldbank.org/indicator/SI.POV.GINI?order=wbapi_data_value_2012+wbapi_data_value+wbapi_data_value-last&sort=asc, last accessed 13 August 2015.
43. Mills: discussion, 25 March 2015.

44. See Milford Bateman, 'Medellín emerges as a Latin American trailblazer for local economic growth', *The Guardian*, 3 April 2012, http://www.theguardian.com/global-development/poverty-matters/2012/apr/03/medellin-trailblazer-local-economic-growth, last accessed 13 August 2015.

45. Mills: discussion, Medellín, March 2015.

46. Mills: discussion, Bogota, March 2015.

47. See Luis Andrade and Andres Cadena, 'Colombia's lesson in economic development', *McKinsey Quarterly*, July 2010.

48. Mills: discussion, Bogotá, March 2015.

49. Mills, Kilcullen and Davis: roundtable discussion with local officials, La Macarena, June 2014.

50. This section draws from 'Colombia' in Greg Mills, *Why States Recover*, Johannesburg and London: Picador and Hurst, 2014 and 2015.

51. This section is based on Mills's visit to the region with the then defence minister and his command team, November 2013.

52. Mills: discussion with Bedoya, La Gabarra, November 2013.

53. These figures were provided by Diana Quintero, Vice Minister of Defence, Bogotá, 26 March 2015.

54. Mills, Kilcullen and Davis: discussion, 27 March 2015.

55. In February 1962 a US Special Warfare team headed by General William P. Yarborough visited Colombia for a follow-on survey to a 1959 security assessment. The resultant policy was instituted as Plan Lazo in 1962 and called for both military operations and civic action programmes in areas affected by Liberal-Conservative violence and growing guerrilla unrest. Following Yarborough's recommendations, the Colombian military recruited civilians into 'civil defense' groups working alongside the military in its counterinsurgency campaign, as well as in civilian intelligence networks to gather information on guerrilla activity. See Grace Livingstone, *Inside Colombia: Drugs, Democracy, and War*, New Jersey: Rutgers University Press, 2004.

56. The US support for helicopters in particular to Colombia began much earlier than Plan Colombia. At the same time, the US provided nearly $1.3 billion to Colombia through Plan Colombia's non-military aid: Alternative Development ($500m); assistance to Internally Displaced Persons ($247m); Demobilisation and Reintegration ($44m); Democracy and Human Rights ($158m); and the Promotion of the Rule of Law ($238m). Figures from 'Plan Colombia', Wikipedia, http://en.wikipedia.org/wiki/Plan_Colombia, last accessed 14 August 2015.

57. Mills: discussion, 27 March 2015.

58. Connor Paige, 'Obama proposes reducing US aid funds to Colombia in 2015', Colombia Reports, 4 March 2014, http://colombiareports.co/us-president-proposes-reducing-aid-funds-colombia-2015/, last accessed 14 August 2015.

59. The wealth tax is imposed on personal holdings of more than US$5 million, at 1.5 per cent over four years.

60. 'President Bush, President Uribe of Colombia Discuss Terrorism and Security', White House Archives, http://georgewbush-whitehouse.archives.gov/news/releases/2005/08/text/20050804–2.html, last accessed 14 August 2015.

61. 'Trade summary for Colombia', World Bank, http://wits.worldbank.org/countrysnapshot/COL/textview, last accessed 14 August 2015.

62. Mills: discussion, Bogota, March 2015.

63. Mills: interview, US Embassy, March 2015.

64. 'Colombia Crude Oil Production', Trading Economics, http://www.tradingeconomics.com/colombia/crude-oil-production, last accessed 25 August 2015.

65. Nick Cunningham, 'Amid Declining Latin American Output, Colombian Oil is Booming', OilPrice, 10 December 2013, http://oilprice.com/Geopolitics/South-America/Amid-Declining-Latin-American-Output-Colombian-Oil-is-Booming.html, last accessed 25 August 2015.

66. Nat Smith, '3 factors in Colombia's 2015 economy', Colombia Reports, 9 January 2015, http://colombiareports.com/3-key-factors-colombias-2015-economy/, last accessed 25 August 2015.

67. Nat Smith, 'Colombia's peso hits 5-year low against dollar', 7 January 2015, http://colombiareports.com/colombias-peso-hits-3-year-low-dollar-now/, last accessed 25 August 2015.

68. IMF, 'Drop in Oil Prices Poses Challenges for Colombia', http://www.imf.org/external/pubs/ft/survey/so/2015/car060815a.htm, last accessed 25 August 2015.

69. Robinson, 'Colombia: Another 100 Years of Solitude?', p. 48.

70. For a summary of these reforms, see 'Colombia's economic outlook is strong, but deep challenges remain, OECD says', OECD, 31 January 2013, http://www.oecd.org/newsroom/colombiaseconomicoutlookisstrongbutdeepchallengesremainoecdsays.htm, last accessed 14 August 2015.

71. Christian Voelkel, 'Three Reasons why Colombia's Land Reform Deal is Significant', International Crisis Group blog, 28 May 2013, http://blog.crisisgroup.org/latin-america/2013/05/28/three-reasons-why-colombias-land-reform-deal-is-significant, last accessed 14 August 2015.

72. Mills: discussion, El Nogal, Bogotá, 25 March 2015.

73. Tom Feiling, *Short Walks from Bogotá*, London: Penguin, 2013, p. 156.

74. 'Coca In The Andes', The White House, Office of National Drug Control Policy, https://www.whitehouse.gov/ondcp/targeting-cocaine-at-the-source, last accessed at 14 August 2015.

75. Kilcullen: telephone interview with Colombian security analyst, Bogota, 22 August 2015.

76. Jeremy McDermott, 'Is Colombia Again the World's Top Cocaine Producer?', InSightCrime, 6 May 2015, http://www.insightcrime.org/news-analysis/colombia-again-world-top-cocaine-producer, last accessed 14 August 2015.

77. Oliver Sheldon, 'Colombia has 10th highest homicide rate in the world: UN',

Columbia Reports, 14 April 2014, http://colombiareports.co/colombia-10th-highest-homicide-rate-world-un/, last accessed 14 August 2015.

78. 'South African Defence Review 2014', http://www.gov.za/sites/www.gov.za/files/dfencereview_2014.pdf, last accessed 14 August 2015.

5. FARC'S TRANSFORMATION: THE COMBINATION OF ALL FORMS OF STRUGGLE

1. By contrast, M-19 enjoyed a significant amount of sympathy among the urban population of Colombia and they were fairly politically savvy, winning a number of local and national elections.

2. Interview, Bogotá, 6 October 2014.

3. 'The World's 10 Richest Terrorist Organizations', Forbes, http://www.forbes.com/pictures/ghki45efh/3-farc-annual-turnover-600-million/, last accessed 15 August 2015.

4. For example, see Delegación de Paz de las FARC-EP, 'Boletín de prensa No. 60: Hablar de irreversibilidad no es conveniente para el proceso', 23 March 2015, http://www.pazfarc-ep.org/index.php/noticias-comunicados-documentos-farc-ep/boletin-prensa/2561-hablar-de-irreversibilidad-no-es-conveniente-para-el-proceso, accessed on 12 April 2015.

5. The best discussion of this is found in Eduardo Pizarro Leongomez, *Las FARC, 1949–1966: del al Autodefensa a la Combinacion de Todas Las Formas de Lucha*, Tercer Mundo Editores, 1991.

6. Jacobo Arenas, 'La Tregua: Problemas de la Guerra y la Paz' in FARC-EP, *Historia de las FARC*, document captured from the computer of Urias Cuellar of FARC's Juan Jose Rondon Column, killed during Operation 7 de Agosto in 2001.

7. As discussed in Chapters 1 and 3, UP was an umbrella party created by FARC that brought the guerrillas, the PCC and some other leftist organisations together in the 1980s to participate in legal electoral politics before disarming as part of the peace process with President Belisario Betancur. The idea of the government and Colombian society was that their electoral success would give FARC the confidence to turn in their weapons and demobilise. FARC never saw the UP as an alternative to armed struggle, but rather as a complement, and FARC's enemies, particularly the drug cartels and paramilitaries, targeted the UP as the soft underbelly of the guerrillas. Many believe that elements of the security forces participated or provided information to the paramilitaries and drug traffickers. Few guerrillas were actually killed, but the PCC was savaged, paying the price for FARC's intransigence towards demobilisation and their attempts to compete with the cartels over drug trafficking areas. See Steven Dudley, *Walking Ghosts: Murder and Guerilla Politics in Colombia*, New York: Taylor & Francis, 2003.

8. Ibid., and see 'Editorial: Renace la Unión Patriótica', *El Tiempo*, 11 July 2013, http://

www.eltiempo.com/archivo/documento/CMS-12924130, last accessed 15 August 2015.

9. Thomas A. Marks, *Colombian Army Adaptation to FARC Insurgency*, Carlisle, PA: Strategic Studies Institute, 2002, p. 20.

10. For a discussion of this classic approach, see Philip Selznick, *The Organizational Weapon: A Study of Bolshevik Strategy and Tactics*, Santa Monica, CA: Rand Corporation, 1952, p. 66–70

11. See 'La infiltrada', *Semana*, 11 October 2007, http://www.semana.com/nacion/articulo/la-infiltrada/89436-3, last accessed 15 August 2015, and 'Alias "Mateo", el infiltrado de las Farc en EPM nunca despertó sospechas', *El Tiempo*, 10 August 2006, http://www.eltiempo.com/archivo/documento/CMS-3110342, last accessed 15 August 2015.

12. For example see 'Inteligencia militar, la marca de Freddy Padilla de León', *Semana*, 27 July 2010, http://www.semana.com/nacion/articulo/inteligencia-militar-marca-freddy-padilla-leon/119806-3, last accessed 15 August 2015.

13. Alfonso Cano, 'Plan Renacer Revolucionario de las Masas', 16 August 2008. In this document, Cano mentioned 'President Chávez', 'Chacin' and 'Senator Piedad'. It is clear he was referring to President Hugo Chávez of Venezuela, Ramon Rodriguez Chacin, then Interior Minister of Venezuela, and Senator Piedad Cordoba of Colombia.

14. Ibid.

15. Available in several languages at the Albert Einstein Institution, http://www.aeinstein.org/free-publications/, last accessed 15 August 2015.

16. 'Cocaleros' is a generic term used across the Andes to refer to coca growers. In Bolivia (and in this book) the term refers to the organisation known as 'Las seis federaciones del trópico de Cochabamba' or the 'Six Federations of the Cochabamba Tropics', organised along union guidelines to advocate for cocalero rights. Subsequently, the federations created the political party, MAS, to participate in electoral politics. In this context, 'Cocalero', 'Federations', and 'MAS' all refer to different aspects of the same group.

17. See previous footnote.

18. One example of many is an interview between Carlos Aznarez and Iván Márquez in Havana, 27 February 2015: 'Comandante Iván Márquez: "Que nadie tenga dudas: nuestro propósito es el socialismo"', Resumen, http://www.resumenlatino-americano.org/2015/02/27/comandante-ivan-marquez-que-nadie-tenga-dudas-nuestro-proposito-es-el-socialismo/, last accessed 15 August 2015.

19. Spencer: private discussion, 2012.

20. In May 2015, President Juan Manuel Santos ordered the end of spraying operations under the pretext of glyphosate's health risks, but many suspect that this was in deference to FARC. See 'Santos ordena suspender fumigación de cultivos ilícitos con glifosato', *El Espectador*, 9 May 2015, http://www.elespectador.com/noti-

cias/judicial/santos-ordena-suspender-fumigacion-de-cultivos-ilicitos-articulo-559592, last accessed 15 August 2015.

21. Government and FARC negotiating commissions in Havana, *Borrador Conjunto: Solucion al Problema de las Drogas Ilicitas*, 16 May 2014.

22. For example see https://resistencia-colombia.org/index.php/dialogos-por-la-paz/comunicados/3467-otro-grito-popular-de-cambio-foro-social-urbano-alternativo-y-popular, last accessed 24 August 2015.

23. Carlos Antonio Lozada, 'Paz O Guerra, El Falso Dilema', Diálogos de Paz, 18 April 2015, https://www.pazfarc-ep.org/index.php/component/k2/2626-paz-o-guerra,-el-falso-dilema, last accessed 15 August 2015.

24. Government and FARC negotiating commissions in Havana, *Borrador Conjunto: Participacion Politica: Apertura democratica para construir la paz*, 6 November 2013.

25. Movimiento Bolivariano Suroccidente de Colombia, http://www.mbsuroccidentedecolombia.org/inicio/manuel-marulanda.html, last accessed 15 August 2015; and 'Farc dicen que no estarán ni un día en la cárcel si se logra la paz', *El Espectador*, 27 March 2013, http://www.elespectador.com/noticias/paz/farc-dicen-no-estaran-ni-un-dia-carcel-si-se-logra-paz-articulo-412750, last accessed 15 August 2015.

26. 'False Positives' is the term used in Colombia to refer to civilians allegedly 'murdered' by unscrupulous individual members of the military who were falsely presented as guerrillas killed in combat. This practice seems to have been motivated by the high premium some military leaders placed on body count and the medals, promotions, plum assignments and rewards that were given to men who produced the highest counts. Several hundred officers and men are currently serving jail terms for this practice.

27. Government and FARC negotiating commissions in Havana, *Borrador Conjunto: Participacion Politica*.

28. Government and FARC negotiating commissions in Havana, *Borrador Conjunto: Hacia un Nuevo Campo Colombiano: Reforma Rural Integral*, 6 June 2013.

29. Ivan Marquez, speech at the opening ceremony of the peace talks in Oslo, Norway, 18 October 2012.

30. 'Farc divulgan plan para una Asamblea Constituyente que selle la paz', El Espectador, 20 December 2013, http://www.elespectador.com/noticias/paz/farc-divulgan-plan-una-asamblea-constituyente-selle-paz-articulo-465356, last accessed 15 August 2015.

31. Delegación de Paz de las FARC-EP, 'Desmilitarizacion de la Sociedad y Reforma de las Fuerzas Militares y de Policía', 7 February 2015, http://www.pazfarc-ep.org/index.php/noticias-comunicados-documentos-farc-ep/boletin-prensa/2442-desmilitarizacion-de-la-sociedad-y-reforma-de-las-fuerzas-militares-y-de-policia, and, http://www.pazfarc-ep.org/index.php/noticias-comunicados-documentos-farc-ep/delegacion-de-paz-farc-ep/2441-propuesta-sobre-provision-de-garantias-reales-y-materiales-de-no-repeticion, last accessed 15 August 2015.

32. Delegación de Paz de las FARC-EP, 'Que se Abran los Archivos pa que se Sepa la Verdad', 19 April 2015, http://www.pazfarc-ep.org/index.php/noticias-comunicados-documentos-farc-ep/delegacion-de-paz-farc-ep/2627-que-se-abran-los-archivos-para-que-se-sepa-la-verdad, last accessed 15 August 2015.

6. COLOMBIA IN COMPARATIVE CONTEXT

1. www.globalhumanitarianassitance.org/country profile/colombia, last accessed 19 August 2015
2. Perhaps the most noteworthy reform was the creation of a Federal War Executive Council supported by state war executive councils and district committees in April 1950, creating an integrated national command and control structure.
3. Karl Hack, 'The Malayan emergency as counter-insurgency paradigm', *Journal of Strategic Studies*, 32(3), pp. 383–414.
4. Field Marshal Sir William Slim, *Defeat Into Victory*, London: Cassell, 1972, p. 147.
5. Letter sent by Field Marshall Montgomery, CIGS, to the Colonial Secretary, Oliver Lyttelton, 23 December 1951.
6. Ong Weichong and Kumar Ramakrishna, 'The Second Emergency (1968–1989): A Reassessment of CPM's Armed Revolution', *RSIS Commentaries*, No 191/2013, 10 October 2013.
7. For a comparative analysis of FARC and Taliban see: Antonio Giustozzi and Francisco Gutierrez 'Networks and armies: Structuring rebellion in Colombia and Afghanistan', *Studies in Conflict & Terrorism*, vol. 33 n. 9, pp. 836–53, September 2010
8. Michael Burleigh, *Small Wars, Far Away Places: The Genesis of the Modern World 1945–65*, London: Macmillan, 2013, p. 184.
9. For more information on this period see Rempe, 'Guerrillas, Bandits, and Independent Republics: US Counterinsurgency Efforts in Colombia 1959–1965', *Small Wars and Insurgencies*, Vol. 6, No. 3 (Winter 1995), pp. 304–327.
10. David Spencer, *Colombia's Road to Recovery: Security and Governance 1982–2010*. Washington: National Defense University, p. 40–41.
11. Davis: interview with Vice President Francisco Santos, Bogotá, 19 March 2015.
12. Davis, Kilcullen and Mills: interview with President Álvaro Uribe, Bogotá, 26 March 2015.
13. John Coates, *Suppressing Insurgency: Analysis of the Malayan Emergency 1948–1954*, Westernview Press, 1992, p. 111.
14. 'Personality Profile: Gerald Templer', *Pointer: Journal of the Singapore Armed Forces*, 2003, Vol 29, No 4.
15. http://www.thefreelibrary.com/Hypocrite+Politician+Ashraf+Ghani+Ahmadzai+and+Know+ Killer+Abdul...-a0345244104, last accessed 19 August 2015
16. 'Ex-finance minister Ghani bullish as Afghan election race begins', AFP Kabul, 16 October 2013.

17. These figures were supplied by the Ministry of Defence of Colombia, May 2015.
18. Study conducted by Headquarters' Regional Command (South) December 2009.
19. http://www.theguardian.com/law/2015/apr/21/malaya-inquiry-batang-kali-massacre-supreme-court, last accessed 19 August 2015
20. Dolf Sternberger, 'Legitimacy' in *The International Encyclopedia of the Social Sciences*, David L. Sills and Robert K. Merton (eds), New York: Macmillan, 1968, p. 244.
21. Amnesty International, 'U.S. Policy in Colombia', www.amnestyusa.org/our-work/countries/americas/colombia/us-policy-in-colombia, last accessed 16 August 2015.
22. Winifred Tate, *Counting the Dead: the Culture of Politics of Human Rights Activism in Colombia*, Oakland: University of California Press, 2007, p. 64.
23. 'Statement by Professor Philip Alston, UN Special Rapporteur on extrajudicial executions—Mission to Colombia 8–18 June 2009', United Nations Office of the High Commissioner for Human Rights, www.ohchr.org/EN/NewsEvents/Pages/DisplayNews.aspx?NewsID=9219&LangID=E, last accessed 16 August 2015.
24. Chris Kraul, 'In Colombia 6 Sentenced in 'false positives' death scheme', *Los Angeles Times*, 14 June 2012.
25. 'Annual report of the United Nations High Commissioner for Human Rights', Annex, 'Report of the United Nations High Commissioner for Human Rights on the situation of human rights in Colombia', 7 January 2013, http://www.ohchr.org/Documents/HRBodies/HRCouncil/RegularSession/Session22/A-HRC-22–17-Add3_English.pdf, last accessed 16 August 2015.
26. Human Rights Watch, 'World Report 2014: Colombia', https://www.hrw.org/world-report/2014/country-chapters/colombia, last accessed 16 August 2015.
27. Álvaro Uribe Vélez, *No Lost Causes*, London: Celebra, 2012, p. 155.

CONCLUSION: PROSPECTS FOR PEACE AND WIDER IMPLICATIONS

1. The section on Buenaventura is based on two field visits by Kilcullen and Mills in June 2014, one by Davis in June 2014, and another by Kilcullen in November 2014.
2. As discussed in Chapter 2, this did not occur through a coup d'etat as in other regional countries, but rather by petition of the elites of the Liberal and Conservative parties, who could not come to an agreement to end the partisan bloodletting of 'La Violencia'. When the political parties tired of the military government four years later, they asked General Rojas Pinilla to leave, which he did without a fight. In this sense, the short military government in Colombia was completely unique in Latin America.
3. See for example Adrian D. Saville and Lyal White, 'Ensuring that Africa keeps rising: The economic integration imperative', *South African Journal of International Affairs*, 22:1, 2015 pp. 1–21, available at http://www.tandfonline.com/doi/pdf/10.1080/10220461.2015.1023342, last accessed 19 August 2015.

4. World Bank, http://data.worldbank.org/indicator/IT.NET.USER.P2, last accessed 19 August 2015

5. World Bank, http://data.worldbank.org/indicator/IT.CEL.SETS.P2?page=1, last accessed 19 August 2015

6. Luke Villapaz, 'Latin American Countries Are The Happiest In The World, Gallup Poll Finds', *International Business Times*, 22 March 2015, http://www.ibtimes.com/latin-american-countries-are-happiest-world-gallup-poll-finds-1855110,   last accessed 16 August 2015.

7. Davis and Mills: interview, Bogotá, March 2015.

8. For a detailed analysis supporting this argument, drawing on an overview of ninety-eight insurgencies since 1945, see Erin M. Simpson, *The Perils of Third-Party Counterinsurgency Campaigns*, Cambridge, Mass: Unpublished Ph.D. dissertation, Harvard University School of Government, 2010, online at http://gradworks.umi.com/34/35/3435578.html, last accessed 24 August 2015.

9. FARC has approval ratings of around 2 per cent. See, for example, Alfredo Rangel, 'The FARC's escalating demands; ongoing attacks and intransigence demonstrate that it doesn't really want peace', *Media in the Americas: Threats to Free Speech*, Fall 2013, at http://www.americasquarterly.org/content/farcs-escalating-demands-ongoing-attacks-and-intransigence-demonstrate-it-doesnt-really, last accessed 26 August 2015. As a comparison, in June 2015, on the back of fresh violence, a regular Gallup poll showed a sharp drop in Colombians' faith in the peace process. Only 33 per cent of those polled—the lowest ever—believed the Havana talks would result in the end of the armed conflict. For the first time since 2003, more respondents (46 per cent) favoured 'no dialogue and try to defeat them militarily' over 'insist on dialogues until a peace accord is reached' (45 per cent) as 'the best option to solve the guerrilla problem in Colombia'. See 'Peace Timeline 2015', Colombia Peace, http://colombiapeace.org/timeline2015/, last accessed 26 August 2015.

INDEX

INDEX

Algerian War (1954–62) 13, 62, 149
All Arms Commando Course 59
Allied Intelligence Bureau 152
Alston, Philip 172
'alternative media' 145
Amazonas Department, Colombia 45
American Civil War (1861–65) 186
Anglo American xxi
AngloGold Ashanti 92, 102, 121
Antioquia Department, Colombia xxv,
    27, 33, 42, 90, 92, 94, 102, 105, 118,
    164, 199
    Medellín xxv, xxxiii, 19, 74, 90–5,
        105–6, 118, 119, 175, 180, 199
    Paramillo Massif 42
    Pavorando 33
    Santa Domingo 56
    Tarazá 28
Apolo JTF 42
Arab Revolt (1936) 156
Arauca Department, Colombia 38, 42
Araújo, Consuelo 35
ARENA (Alianza Republicana Nacio-
    nalista) 186
Arenas, Jacobo 28, 29
Argentina 39, 99
Arnott, Anthony 196, 199
Arpía helicopters 54, 55
assassinations 1, 28, 74
Attorney General's Office (Fiscalia)
    xxxiii
AUC (Autodefensas Unidas de Co-
    lombia) xxv, 8, 10, 83, 95
Australia 152, 155, 159, 160, 199
Australian National University, Can-
    berra 155
autodefensas 82, 170, 197; see also
    AUC, paramilitaries
Autodefensas Campesinas de Córdoba
    y Urabá 82
autonomous zones ('independent
    republics') xxvii, 4–5, 23, 85–6, 191
Aviación del Ejército 49

BACOA (Batallón de Commandos)
    58, 59

BACRIM (Bandas Criminales) xxxiii,
    xxxiv, 2, 7, 11–12, 22, 40, 46, 82–3,
    84, 85, 87, 106, 145, 170, 188
Balkans 186
Banana Wars 69
bananas 19, 69
Bancolombia 92
banking 92, 99
BAOEA (Batallón de Operaciones
    Especiales de Aviación) 59
Barco, Virgilio 25, 28
Barrancabermeja, Santander 20
Barrancominas, Guainía 35
Barranquilla, Atlántico 19, 36, 94
Barrero, Leonardo 111
barrios 90, 92, 180
Basque Country 39
Batang Kali massacre (1948) 171
Bavaria Breweries 101
Bedoya, Jorge Enrique 110
Beira Mar, Fernandinho 35
Bell 212 helicopters 55
Bell UH-1 helicopters 53, 55, 56
Berlin Wall 29, 157, 177
Betancourt, Ingrid 36, 39, 174
Betancur, Belisario 18, 24, 28, 78, 204
'black budget' 51
BlackHawk helicopters xxvii, 9, 54, 55,
    110
black-market currency exchange 72
blackmail 74
blanco y negro ('white and black') 180
bloc-level operations 30–1, 73, 76
Bogotá, Cundinamarca xxii, xxv, xxxiii,
    8, 18, 20, 22, 27, 31, 34–7, 58, 67,
    74, 78, 90, 92, 94–5, 103, 108, 119,
    128, 133, 140, 174, 182, 187
    El Dorado International Airport 182
    El Nogal 37
    Palace of Justice 58
    Soacha 172
Bogotázo (1948) 18
Bojayá, Chocó 70
Boko Haram 15, 121, 180
Bolívar, Simón 3, 59, 140

212

INDEX

INDEX

INDEX

INDEX

Integral Action 80
Integrated Illicit Crop Monitoring
System (SIMCI) 35
intelligence 49, 50–3
Interim Afghan Administration 161
Internal Security Act (Malaysia) 157–9
International Conference on Afghani-
stan (2010) 160, 169
International Criminal Court xxxvi
international humanitarian law xxxii,
xxxiv, 29, 170
International Monetary Fund (IMF)
47, 96, 116
Internet 144, 182
intra-state conflicts 13, 77
Inversiones Argos 92
Inversura 92
investment 10, 20, 31, 32, 91, 98–103,
115–17, 173, 179, 181, 182, 186
IRA (Irish Republican Army) 33, 35,
68, 74
Iran 68
Iraq War (2003–11) 13, 46, 56, 60,
149, 160, 173, 184, 197
Ireland 33, 35, 46, 68, 74, 176, 186,
197
ISAF (International Security Assis-
tance Force) 159–60, 168, 169, 173
Israel 48, 51, 52, 53, 55, 112, 196
Italy 39

JACs (Juntas de Acción Comunal)
82, 86
Japan 91, 152, 153
Jara, Alan 35
Jaramillo, Sergio 43
Johannesburg, South Africa xxi
Joint Operational Intelligence Com-
mittee (JOEC) 52
Joint Special Operations Command
(CCOES) 48, 58, 59
Joint Task Forces (JTFs) xxvii, xxxiii,
xxxv, 11, 38, 42, 43, 73–4, 80, 81,
87, 108–9, 188
Omega xxvii, 38, 42, 109

Jojoy, Mono (Jorge Briceño Suarez) 37,
40, 48, 56, 60, 76, 108, 109
Jose Maria Cordoba Bloc 30
judicial police units ('Groic') xxxiii
Jurado, Chocó 192

Kabul, Afghanistan 159, 177
Kabul University 161
Karzai, Hamid 161, 165, 166, 173, 183
Kenya 13, 176
Kfir 48, 53
Khorasan Province, Afghanistan 178
kidnappings xxv, xxvi, xxix, 1, 6, 12, 23,
27, 35, 36, 69–71, 74, 75, 103, 133,
180, 181
Korea 91
1950–53 Korean War 50, 164, 170

Lagos, Nigeria 180
Lanceros 59
land reform 23, 104, 118–19, 186
Land Restitution and Victims' Law
(2011) 13, 104–5
Law 001 on land reform 23
League of Nations 46
Lean, David 110
Leticia, Amazonas 45
LGBT (lesbian, gay, bisexual and trans-
gender) community 139, 144
Liberal Party of Colombia 3, 4, 5,
18–19, 22, 23, 25, 96, 97, 102, 202,
208
Liberation Theology 5
Libya 7
Line J, Medellín Metrocable 90
Lockhart, Rob 156
London, England 156, 160, 169
Londoño Echeverri, Rodrigo (Timo-
chenko) 110, 134
Londoño, Santiago 105, 106
Lord's Resistance Army 7
Loya Jirga 165
Lunes de Recompensa 51
Lyttelton, Oliver 155, 162, 164, 166,
176

222

INDEX